Customer-driven manufacturing

JOIN US ON THE INTERNET VIA WWW, GOPHER, FTP OR EMAIL:

WWW: http://www.thomson.com
GOPHER: gopher.thomson.com
FTP: ftp.thomson.com
EMAIL: findit@kiosk.thomson.com

A service of I(T)P®

Customer-driven Manufacturing

Edited by

J.C. Wortmann
Technische Universiteit Eindhoven
The Netherlands,

D.R. Muntslag
Moret, Ernst & Young
Utrecht
The Netherlands

and

P.J.M. Timmermans
CIMRU, University College
Galway
Republic of Ireland

CHAPMAN & HALL
London · Weinheim · New York · Tokyo · Melbourne · Madras

Published by Chapman & Hall, 2–6 Boundary Row, London SE1 8HN, UK

Chapman & Hall, 2–6 Boundary Row, London SE1 8HN, UK

Chapman & Hall GmbH, Pappelallee 3, 69469 Weinheim, Germany

Chapman & Hall USA, 115 Fifth Avenue, New York, NY 10003, USA

Chapman & Hall Japan, ITP-Japan, Kyowa Building, 3F, 2-2-1 Hirakawacho, Chiyoda-ku, Tokyo 102, Japan

Chapman & Hall Australia, 102 Dodds Street, South Melbourne, Victoria 3205, Australia

Chapman & Hall India, R. Seshadri, 32 Second Main Road, CIT East, Madras 600 035, India

First edition 1997

© 1997 Chapman & Hall

Printed in Great Britain by T J Press, Padstow, Cornwall

ISBN 0 412 57030 0

Apart from any fair dealing for the purposes of research or private study, or criticism or review, as permitted under the UK Copyright Designs and Patents Act, 1988, this publication may not be reproduced, stored, or transmitted, in any form or by any means, without the prior permission in writing of the publishers, or in the case of reprographic reproduction only in accordance with the terms of the licences issued by the Copyright Licensing Agency in the UK, or in accordance with the terms of licences issued by the appropriate Reproduction Rights Organization outside the UK. Enquiries concerning reproduction outside the terms stated here should be sent to the publishers at the London address printed on this page.

The publisher makes no representation, express or implied, with regard to the accuracy of the information contained in this book and cannot accept any legal responsibility or liability for any errors or omissions that may be made.

Publisher's note
This book has been prepared from camera ready copy provided by the editors

A catalogue record for this book is available from the British Library

Library of Congress Catalog Card Number: 96-086530

∞ Printed on permanent acid-free text paper, manufactured in accordance with ANSI/NISO Z39.48-1992 and ANSI/NISO Z39.48-1984 (Permanence of Paper).

Contents

List of contributors   xi

Foreword   xiii

1 Introduction   1

    1.1 Customer-driven manufacturing: how and why   1
    1.2 A design-oriented approach   2

Part A BACKGROUND

2 A design-oriented approach   9

    2.1 Introduction   9
    2.2 Design choices and performance indicators linked by models   10
    2.3 The need for explicit modelling   11
    2.4 Types of models   14
    2.5 Model variables   17
    2.6 Summary and conclusion   19

3 Customer-driven manufacturing in a macro-economic perspective   21

    3.1 Introduction   21
    3.2 Mass production and mass customisation v genuine customer orientation?   22
    3.3 Beyond the hype   24
    3.4 The road to improvements   28
    3.5 Conclusions   30

4 Why customer-driven manufacturing   33

    4.1 Introduction   33
    4.2 General development trends in manufacturing   33
    4.3 Global challenges for future production systems   34
    4.4 Economic and technical challenges for future production systems   35
    4.5 Customer-driven manufacturing systems   37
    4.6 Summary   43

5 Describing production situations   45

    5.1 Introduction   45
    5.2 The foundation of a framework for description   47
    5.3 Three extensions to the model   49
    5.4 Conclusion   57

6 A typology of customer-driven manufacturing — 59

- 6.1 Introduction — 59
- 6.2 The typology — 59
- 6.3 The nature of production management in different types of typology — 63
- 6.4 Information systems for one-of-a-kind production — 68
- 6.5 Summary and conclusions — 73

7 Overview of the book — 75

- 7.1 Introduction — 75
- 7.2 Structure of the book — 76

Part B CASES

8 Introduction to the cases — 85

9 Engineer-to-order in product-oriented manufacturing — 89

- 9.1 Introduction and company description — 89
- 9.2 The internal organization (resources view) — 90
- 9.3 The transformation process (workflow view) — 93
- 9.4 Present management of the transformation process — 95
- 9.5 Further analysis of the problem issues — 97

10 Assemble-to-order in product-oriented manufacturing — 105

- 10.1 Introduction — 105
- 10.2 Company description — 105
- 10.3 The range of products — 109
- 10.4 The transformation processes (workflow view) — 111
- 10.5 Contemporary methods of controlling the primary process — 113
- 10.6 Further analysis of the problem issues — 117
- 10.7 Summary — 118

11 Make-to-order in work-flow-oriented manufacturing — 119

- 11.1 Introduction — 119
- 11.2 Company description and situation — 119
- 11.3 Internal organisation (resources view) — 121
- 11.4 The transformation process (workflow view) — 122
- 11.5 The production planning system — 126
- 11.6 Further analysis of the problem issues — 133
- 11.7 Conclusion — 135

12 Engineer-to-order in resource-oriented manufacturing — 137

- 12.1 Introduction and description of the company — 137
- 12.2 The internal organization (resource view) — 139
- 12.3 The transformation process (work-flow view) — 140
- 12.4 Current production planning and control — 148
- 12.5 Further analysis of the problem areas — 151

Part C WORKFLOW AND RESOURCES

13 Introduction to Part C — 155

14 Engineering in customer-driven manufacturing — 159
14.1 Introduction — 159
14.2 The structure of the engineering process — 160
14.3 Design requirements resulting from cost control — 163
14.4 Design requirements resulting from production control — 165

15 Production in CDM — 177
15.1 Introduction — 177
15.2 The resources structure — 180
15.3 Production flow analysis and group technology — 184
15.4 Design considerations with respect to manufacturing technology — 186
15.5 Concluding remarks — 188
Appendix — 190

16 Product Modelling and other functions — 193
16.1 Introduction — 193
16.2 Product models: an overview — 194
16.3 Shortcomings of these product models — 195
16.4 Modelling a range of product variants — 197
16.5 Sales–manufacturing communications — 205
16.6 Summary and conclusions — 207

17 Human resource management in customer-driven manufacturing — 209
17.1 Introduction — 209
17.2 Basic resource management requirements — 211
17.3 Work systems as a meaningful synthesis of resources — 214
17.4 Challenges for resource management of resource-oriented systems — 216

Part D ORGANISATION AND DECISION MAKING

18 Organisation and decision making in customer-driven manufacturing — 223
18.1 Introduction — 223
18.2 Requirements for competitive customer-driven manufacturing systems — 223
18.3 The GRAI model — 224
18.4 Modelling engineering activities using the GRAI model — 231
18.5 CDM systems: a reference model for management structure, based on GRAI model — 233
18.6 Conclusion — 237

19 Production control in workflow-oriented make-to-order forms — 241
19.1 Introduction — 241
19.2 Production control design principles — 244
19.3 Designing the production process structure — 249
19.4 Production control in more detail: the fine-paper factory – proposals for improvement — 251

20 Production control in product-oriented assemble-to-order manufacturing — 257

20.1 Introduction — 257
20.2 Customer-driven assembly of systems from subsystem at Medicom — 258
20.3 Customer-driven assembly of systems and subsystems from components at Medicom — 263
20.4 Customer-driven assembly in repetitive manufacturing — 265
20.5 conclusion — 265

21 Production control in engineer-to-order firms — 267

21.1 Introduction — 267
21.2 Production control design principles — 267
21.3 Structuring the transformation process — 269
21.4 Developing the production control framework — 271
21.5 developing the decision structure — 274

22 Management of tendering and engineering — 281

22.1 Introduction — 281
22.2 The control system for customer-driven engineering — 281

23 Production Activity Control — 289

23.1 Introduction — 289
23.2 Work load control — 291
23.3 Scheduling: capacity allocation and detailed work order planning — 294
23.4 Dispatching — 296
23.5 Conclusions — 301

24 Customer and supplier relations — 303

24.1 Introduction — 303
24.2 Manufacturer–customer relations — 303
24.3 Manufacturer, supplier and subcontractor relations — 309
24.4 Summary — 313

Part E INFORMATION TECHNOLOGY

25 Introduction to Part E: Information Technology — 317

26 Generic Product Modelling & Information Technology — 319

26.1 Introduction — 319
26.2 Data model for generic bills-of-material — 319
26.3 Generic bill-of-material application — 322
26.4 Developing product families — 324
26.5 Design rules — 326
26.6 Conclusions — 331

27 Standard software packages for business information systems in customer-driven manufacturing — 333

27.1 An architecture for business information systems — 333
27.2 Information systems for product-oriented make-to-stock production — 336
27.3 Information systems for engineer-to-order production — 344
27.4 Conclusion — 356

28 Document management in customer-driven manufacturing — 357

- 28.1 Introduction — 357
- 28.2 Trends in document management — 358
- 28.3 Document management activities — 359
- 28.4 Benefits of document management in customer-driven manufacturing — 365
- 28.5 The role of a document management system — 371

29 IT-support of customer-driven engineering management — 375

- 29.1 Introduction — 375
- 29.2 The state-independent data model — 376
- 29.3 The state-dependent data model — 380
- 29.4 The relationships between the data models — 382

30 Production Unit Control — 385

Part F MODELS OF THE FOF WORKBENCH

31 Introduction to the FOF Workbench — 397

- 31.1 Introduction — 397

32 An intelligent storage and retrieval for design choices and performance indicators — 401

- 32.1 Introduction — 401
- 32.2 Rembrandt — 402
- 32.3 Software and hardware platform — 402
- 32.4 Using Rembrandt — 403
- 32.5 Graphical presentation of the network — 405
- 32.6 Conclusion — 406

33 A simulation model for human resource management in customer-driven manufacturing — 407

- 33.1 Motivation — 407
- 33.2 Basic approach — 410
- 33.3 Technical structure of the simulation model — 413

34 Group Design — 417

- 34.1 Introduction — 417
- 34.2 Describing capacities and capabilities of working groups — 418
- 34.3 Design choices — 421
- 34.4 Performance indicators — 422
- 34.5 Relating design choices and performance indicators — 423
- 34.6 Conclusion — 425

35 Departmental coordination model — 427

- 35.1 Business problems addressed by the model — 427
- 35.2 Theoretical background and available theories — 428
- 35.3 Relevant DC PI relationships — 428
- 35.4 Description of the implemented model — 433

36 Interdepartmental coordination — 437
  36.1 Situation in the factory — 438
  36.2 Design choices — 438
  36.3 Evaluation — 442

37 XBE in design of customer-driven manufacturing systems — 445
  37.1 Introduction — 445
  37.2 Problem area — 445
  37.3 Theoretical background — 446
  37.4 The toolbox — 448
  37.5 Description of the model — 449

Contributors

Chapter	Author(s)	Institute
1	Hans Wortmann	TUE
2	Rob Kwikkers	IPL
3	Eero Eloranta	HUT
4	Bernd Hirsch / Klaus-Dieter Thoben	BIBA
5	Peter Falser	DTH
6	Hans Wortmann	TUE
7	Patrick Timmermans	TUE
8	Peter Falster / Hans Wortmann	DTH/TUE
9	Dennis Muntslag	TUE
10	Freek Erens	TUE
11	Fons van Aert / Hans Wortmann	TUE
12	Klaus-Dieter Thoben	BIBA
13	Asbjorn Rolstadas	SINTEF
14	Dennis Muntslag	TUE
15	Patrick Timmermans / Hans Wortmann	TUE
16	Freek Erens	TUE
17	Bernd Hamacher	BIBA
18	Guy Domeingts	GRAI
19	Fons van Aert / Hans Wortmann	TUE
20	Hans Wortmann	TUE
21	Will Bertrand / Dennis Muntslag	TUE
22	Dennis Muntslag	TUE
23	Patrick Timmermans / Hans Wortmann	TUE
24	Klaus-Dieter Thoben	BIBA
25	Jimmie Browne / Hans Wortmann	CIMRU/TUE
26	Freek Erens	TUE
27	Hans Wortmann	TUE
28	Pierre Breuls	TUE
29	Dennis Muntslag / Hans Wortmann	TUE
30	Jimmie Browne / Hans Wortmann	CIMRU/TUE
31	Eero Eloranta	HUT
32	Juho Nikkola	HUT
33	Bernd Hamacher	BIBA
34	Pierre Breuls	TUE
35	Francois Marcotte / Patrick Timmermans	GRAI/TUE
36	Francois Marcotte / Patrick Timmermans	GRAI/TUE
37	Juho Nikkola	HUT

Foreword

Customer Driven Manufacturing is a book with a special history. In spring 1988 the European R&D programme for Information Technology, ESPRIT, was extended to include Basic Research Actions. At several meetings with Bernd Hirsch, Guy Doumeingts, Eero Eloranta, Peter Falster, Jimmie Browne and others, I was encouraged to write a proposal for basic research into the nature of the Factory of the Future (FOF), a proposal that would not have been realised without the help of Bernd Hirsch and members of BIBA. The proposal clearly presumed that customer driven manufacturing was a key element in future manufacturing systems and was selected for a contract by the European Commission.

The FOF project started in early 1989 and consisted of a consortium of seven research institutes, viz.:

- Bremen Institute of Manufacturing Automation, Fed. Rep. of Germany (BIBA)
- CIM Research Unit, University of Galway, Ireland (CIMRU)
- Dansk Technische Hochskole, Lyngby, Denmark (DTH)
- Eindhoven University of Technology, The Netherlands (TUE)
- Grai Laboratories, Univ. Bordeaux I, France (GRAI)
- Helsinki University of Technology, Finland (HUT)
- SINTEF, Institute for Technology, Trondheim, Norway (SINTEF).

Technical management was provided by TNO-IPL (Rob Kwikkers). Although the work was performed by research institutes, there was much interaction with industrial companies, notably the Netherlands Machine Factory, Alkmaar, and Bremen Vulcan shipyard in Bremen.

FOF was a peculiar project. On the one hand, it was a typical basic research effort with many uncertainties and problems not clearly defined. Due to different scientific cultures, some people commenced with formalisations of problems which others did not even recognise as problems. On the other hand, it created a strong common view that customer driven manufacturing is challenging, human-oriented, relevant to the future, and to some extent enabled by modern IT.

Following an empirical study carried out under the responsibility of Guy Doumeingts, the first theoretical framework was constructed through the personal effort of Peter Falster, and —at a later stage — Asbjorn Rolstadås. In essence, this framework is still present in Chapter 5 of this book. Under the inspiring leadership of Eero Eloranta, the FOF consortium built a number of computerised models, some of which are presented in the last part of this book. Jim Browne assumed responsibility for placing all these models in a single framework. This framework and the models were used for the rapid prototyping of alternatives to existing practical situations and were presented at a conference in Bremen on one-of-a-kind production organised by IFIP 5.7. The conference chairman was Bernd Hirsch.

FOF gave much inspiration to each of the participants and received various favourable reviews. The consortium therefore decided to disseminate the results. On the one hand, it designed a number of courses under the European COMETT programme, which focuses specifically on education. This work is now being performed under the name EFSATT, which is a COMETT contract with PETIME, an established COMETT institution (more specifically, a University-Enterprise Transfer Point or UETP). PETIME is located in Trondheim, Norway. On the other, the FOF consortium initiated this book, which has been sponsored by PETIME.

The book has many authors, as contributions have been included from all the FOF partners as well as several others. However, the editorial team has tried to turn it into an integrated whole. Dennis Muntslag, who was not involved in the FOF project itself but who wrote a thesis on a related project, was the driving force for more than a year. He wrote several chapters on engineering to order and reviewed all the material critically. It was fortunate that he had not been involved in FOF earlier, so that all the material was fresh to him. Patrick Timmermans was the continuous factor in the project, as earlier in FOF. He knows all the authors personally and is very adept at putting his efforts where they are needed. I am much indebted to both these editors for their outstanding work. I hope that many practitioners in customer driven companies will benefit from this book.

Hans Wortmann

1

Introduction

1.1 CUSTOMER DRIVEN MANUFACTURING: HOW AND WHY

Customer driven manufacturing is the key concept for the factory of the future. The markets for consumer goods are nowadays marked by an increase in variety, while at the same time showing steadily decreasing product life-cycles. In addition, tailoring the product to the customer's needs is becoming increasingly important in quality improvement. These trends are resulting in production in small batches, which are driven by customer orders. Detailed arguments for this view, which is fundamental to the whole book, will be given throughout Part A, which comprises chapters 2 to 7.

There is much experience with customer driven manufacturing in the production of capital goods. However, capital goods industry does not always hold a flourishing position. The question is: why not? We believe it is often due to the thoughtless copying of systems designed for mass production.

Customer driven manufacturing requires greater customer satisfaction at lower cost than the competition. How can this be achieved? First of all, by focusing on the customer - as in service organisations. Who should focus on the customer? The company's own employees, of course. Thus, customer driven manufacturing requires motivated and service-oriented employees. Such employees are created by excellent management, especially *human resource* management.

However, service orientation alone on the part of employees will not create a profitable business. The customer should also get high added value at reasonable cost. Speed and quality always contribute to high added value *and* to lower costs. In addition, customer driven manufacturing requires a strategy. This strategy specifies the external market and the internal focus. The focus can be on capabilities and/or work flows and/or products. Chapters 3 and 4 elaborate on strategies for obtaining value-added customer driven manufacturing.

2 Introduction

Although this book focuses on manufacturing, it should be borne in mind that the supply of goods always requires physical distribution. Therefore, customer driven manufacturing implies customer driven distribution. Also, customer driven manufacturing is increasingly sub-contracted, so that a customer driven supply chain emerges. At several points in this book we shall refer explicitly to this supply chain. Sales and purchasing activities are always considered as components of manufacturing business processes.

1.2 A DESIGN-ORIENTED APPROACH

This book is design-oriented. The question of how production systems can be designed or redesigned is always in the background, and often in the foreground, too. However, the nature of the design of production systems requires some discussion.

When a physical artifact is designed, manufactured, repaired or dismantled, there is no discussion about the necessity of complete documentation. From a philosophical point of view it may be debated whether *complete* documentation is possible, but the established engineering disciplines have achieved a large degree of consensus on the meaning of "complete documentation". This is true even for a new discipline like software engineering.

When an organisation is created or changed, the situation is completely opposite. There is no common agreement in organisational theory about its documentation. More specifically, there is no agreement on description languages, either in terms of semantics, or in terms of syntax. It can even be doubted whether organisations should be considered as artifacts or as organisms which live, grow, generate progeny and die. However, although there is no common agreement on the documenting of organisations, there is quite some theory about the functioning of an organisation, about its structure, its systems, its economy, its employees and its contribution to society. But all this theory is fragmented.

This fragmentation shows up very clearly in practice when interfaces between automated subsystems have to be designed. For example, consider the situation in which a CAD system has to be interfaced with an MRP system. Quite often this means that a product design view of "engineering data management" is confronted with a production control view of "engineering data management", and the result is seldom an elegant, integrated whole.

This discussion brings us to another type of fragmentation which should be mentioned, viz. the fragmented description of the work flow through a company. Although many companies have described several flows of work in their procedural handbooks, few have taken the work flow as the basis for their organisational design or for their systems design. This is especially true for customer-order-driven engineering work.

The disadvantage of neglecting the work flow can be observed in many places where computer aided engineering (CAE) has been installed. Usually, CAE systems provide considerable support to individual engineers to improve the quality of their work, but they do not contribute to improving cooperation in a team that has to manage a flow of work. In other words, they help in doing the work right, but not in doing the right work.

Both types of fragmentation are illustrated in Figure 1.1.

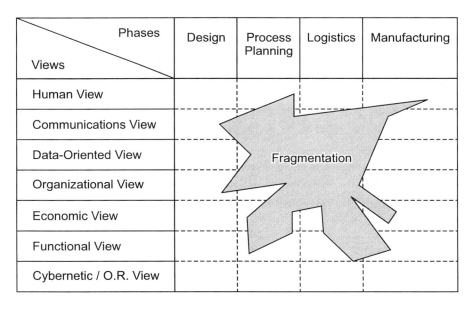

Fig. 1.1 Fragmentation in views and production phases.

This book does not claim to eliminate this fragmentation completely, but it shows the road towards integration. Integration is considered to occur if agreement can be reached on:

- a method of evaluating proposed changes in a production system in terms of effects on performance indicators
- the elements in a production system that need to be described in order to appraise its functioning or manage its operations.

The first issue is called the *design framework*. It is introduced in Chapter 2 of this book. This chapter discusses the idea that design or redesign should focus on design choices that should be related to performance indicators. These design choices are discussed in more detail in subsequent parts of this book. The first discussions - of cases - are found in Part B, while Part C discusses the design of the production environment, Part D production management and Part E the role of information technology. The design framework is illustrated in depth in Part E, where a number of formal models are given as examples illustrating the relationship between design choices and performance indicators.

The second issue is referred to as the *theoretical framework* within this book. It is introduced in Chapter 5 and describes the elements needed when documenting a production system. It shows how the above fragmented views can be compared and combined. This produces a reduction to three complementary views, necessary to describe a production system with the aim of appraising its functioning. These views are:

- the work flow view
- the resource view
- the organisational/decisional view.

All this modelling can be criticised for the fact that it does not recognise one basic point, namely the idea that production systems can only partially be the object of design or redesign and should also partially be the subject of redesign. In other words, *production systems should acquire the ability to change themselves* rather than be changed by some external agent of change. This issue will be discussed in Part E.

Based on the theoretical framework, this first Part elaborates further on the nature of customer driven manufacturing in Chapter 6. It discusses questions like: What is customer driven manufacturing? Is the automotive industry customer driven?

These questions are dealt with by developing a typology of customer driven production in Chapter 6. The typology does not solve all the issues but it at least helps in understanding some of them.

The typology is based on two dimensions. The first dimension is well known and poses the question:

a. Up to which point in the work flow are the production activities driven by the customer order (the customer-order decoupling point)?

The second dimension is less well-known and asks the question:

b. To what extent has the company invested in product development (and/or work flow process development) *in*dependent of the customer order?

The typology is used to differentiate between the cases presented in the next Part. Many design choices also depend on the place in the typology at which a company is positioned.

Finally, Part A ends with Chapter 7, which gives an overview of the remainder of the book.

Part A
BACKGROUND

2

A design-oriented approach

2.1 INTRODUCTION

This chapter presents the "Design Choice - Performance Indicator" (DC-PI) modelling approach. The DC-PI modelling approach supports the design of a production system in a quantified and multidisciplinary manner. It provides the basis for integrating various scientific and engineering disciplines involved in the design and redesign of production systems. As such this modelling approach is one of the main elements of this book.

Customary performance indicators are primarily of a financial nature. For customer driven manufacturing systems especially, performance indicators from various other fields (logistics, quality control, human resource management) are equally important. Traditionally, production systems redesign means different things to different professionals.

In this chapter we will first explain the notion of design choices and performance indicators. We will then see why it is beneficial to have available explicit models which provide links between design choices and performance indicators. Next, we will introduce the DC-PI modelling framework. This framework presents the production system design/redesign issue in various degrees of detail, thus allowing us to explore parts of redesign in-depth while maintaining an overview. After that, we will zoom into the models themselves. We will explore the types of models that can be used to link design choices with performance indicators and indicate the circumstances in which certain types are appropriate. Finally, we will go into the types of variables that play a role in the models and how they can be used in a simulation study to assess the link between design choices and performance indicators. These models are described in more detail in Part F of this book.

2.2 DESIGN CHOICES AND PERFORMANCE INDICATORS LINKED BY MODELS

A designer makes Design Choices
Designing implies choosing between alternatives. A product designer can choose to use metal or plastic, wood or cork as the basic material for his creation. A production system designer can choose to use manually controlled or fully automated machinery. He has many such choices, which are often not duly recognised.

With respect to the planning of work, for example, a production system designer has a wide range of alternatives regarding the level of detail employed. At one extreme he may decide to release only fully detailed drawings to the shop floor with work completely planned. At the other, he may decide to rely on the individual judgement and creativity of the workers on the shop floor. Another example can be taken from production control. A production system designer can choose to use a detailed scheduling system or to use a simple input-output workload control system. A third example relates to human resource management. A human resource manager can choose to base the flexibility of the company on hire/fire and subcontracting strategies or on training multiskilled employees.

When designing and managing a production system, it is necessary to make many such choices. These choices are collectively referred to here as Design Choices.

Performance Indicators are used for assessment
Designers make the design choice that optimises the performance of the production system. To make the decisions, the effects of the alternatives are assessed and the choice resulting in optimum performance is made. The product designer would assess the performance of the product in terms of fitness for use, aesthetics, and material cost. Of course, he employs derived performance indicators such as maintainability for components of the product.

The production system designer would look at the performance of the overall production system in terms of efficiency, flexibility and human empowerment, for example. Naturally, the production system designer employs derived performance indicators for parts of the production system. With respect to work planning, for example, the production system designer might assess the performance in terms of planning process efficiency, the expected quality of the resulting planning and the number of mistakes made.

With respect to the second example, the production control system, the production system designer might assess the efficiency and throughput time that can be achieved with various control methods as relevant performance indicators. As regards human resource management, finally, the designer might look, for example, at the cost of adapting to market needs and the speed at which this adaptation can be effected.

Design Choices are linked to Performance Indicators by Models
The assessment of design choices made by various designers is based on estimations of their effect on various performance indicators. To make this estimation they use a cause-and-effect model. This model links performance to indicators that the designer deems important to the choice he is about to make in the design. For example, the product designer may know that products made of metal are sturdier than those made of cork. The production equipment designer may know that hard-automated systems are more efficient but less flexible than systems using general purpose tools controlled by people. The designer of the work-planning organisation may know that planning by the workers on the floor results in lower process efficiency but fewer mistakes. And so on.

Common to all these assessment processes is that the designer is aware of the link between the design choice and the performance indicator and has some means of predicting the performance (and differences in the performance) when considering various design alternatives. In other words, the production system designer has a model that links design choices to performance indicators.

The models used by the designer may be implicit models where his/her knowledge and experience is applied informally to the choice question, or formal models where captured knowledge and experience are applied. These models are at the core of the DC-PI modelling approach. The next section goes into the reasons why explicit models are important.

2.3 THE NEED FOR EXPLICIT MODELLING

A designer uses models to link design choices to performance indicators. When these models are implicit, they only exist in the designer's mind. If so, they are not described in a manner that is useful to others. When these models are explicit, they can be communicated, which can be appreciated here. However, many design models are based on hidden assumptions or on specific conditions which are context-dependent. Therefore, it is hard to integrate these models into a coherent whole.

Explicit DC-PI models form captured knowledge and experience, described in such a way that it can be shared between scientists and practitioners and employed for a multidisciplinary assessment of practical design choices.

Sharing knowledge and experience

In many engineering disciplines related to manufacturing, developments are so rapid that it is impossible for a single person to assess the practical implications for production system redesign. Therefore, it is absolutely necessary to make knowledge and experience from many sources accessible to other designers in the field. The DC-PI modelling approach provides a framework for sharing new insights. This approach defines a clear format for describing new knowledge: knowledge is made explicit as a relationship between causes and effects. The availability of explicit cause-and-effect models opens up the possibility of including results from other researchers and practitioners to complement one's own knowledge.

The Conceptual Reference Model: DC-PI Network

To obtain an overview of all related factors in a production system design, our reference model connects design choices (DCs) to performance indicators (PIs). For this reason it is called a *connectance model*. The idea is that design choices are changed, and the effect on the system is measured by the performance indicators. The reference model takes the form of a network and is called the DC-PI network.

Design choices could be physical layout, socio-technical structure, functional structure, decision hierarchy and responsibility, or organisation. Alternatives in the design could be functional, process or group layout, hierarchical or autonomous group structure, etc. The aim of modifying design choices is to obtain a better score on the performance indicators deemed relevant to the design decision. Performance indicators could be efficiency, quality of working life, decision stability, etc.

The structure of the DC-PI network is such that it can be maintained if our knowledge of manufacturing systems grows. Part of a possible DC-PI network is presented in Figure 2.1 below. (It will be explained in detail in Chapter 32.)

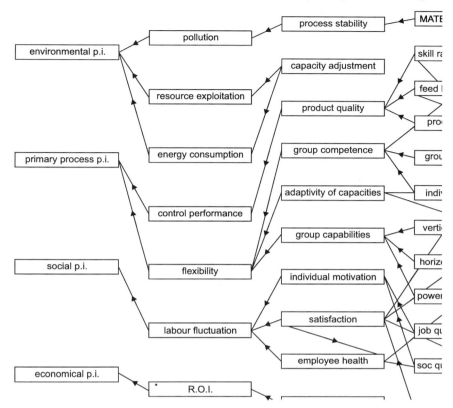

Fig. 2.1 Sample from a DC-PI network

The type of connections implemented in the reference model are:

- Affects (is affected by)
- Is element of (has element)
- Increases (is increased by) / Decreases (is decreased by)
- Complicates (is complicated by) / Simplifies (is simplified by)

Because of the variety of permitted relationships, the mathematical properties and hence the reasoning operations in the reference model are scarce. Even transitivity is limited to just a few types of relationship.

Relationship Models

The conceptual reference model provides an overview. However, other models are needed to obtain more specific insight into details. Therefore, we will introduce in Part F models which focus on a part of the DC-PI network, and provide further detail. This provides a full understanding of a limited area, usually some scientific or engineering discipline. These models are called *relationship models* because they explain or describe the relationships of the conceptual reference model more specifically.

Relationship models are quantified counterparts of connectance models. They are quantified in such a way that when design choices are changed, the effect on the system can be observed (the performance indicators).

Figure 2.2 depicts the position of relationship models between design choices and performance indicators of a full connectance model.

A relationship exists between each design choice and its relevant performance indicators. These relationships, obtained from theory or

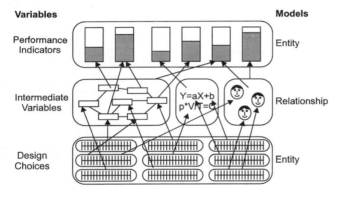

Fig. 2.2 Connectance and Relationship models between DCs and PIs

by empirical methods, are expressed in models (relationship models). It is the task of the modeller to specify these relationship models and, with that, to indicate the relationship between a design choice and its performance indicators.

2.4 TYPES OF MODELS

There are three types of relationship model in the DC-PI modelling framework: Analytic, Dynamic, and Event-Flow. These models differ in the complexity of the DC-PI relationships they capture and therefore in the method of calculation required to evaluate the relationships between Design Choices (DCs) and Performance Indicators (PIs).

Analytic Models
In the simplest case of DC-PI models the relationships are straightforward: for example, return on investment is proportional to profit; the cost of a company's own capacity is proportional to the size of its workforce. Here, the relationships between some DCs and PIs can be evaluated using analytic models. Spreadsheets could be used to evaluate relationships of this type. An example of a DC-PI model for evaluating Human Resource Management strategies for short-term optimisation in a spreadsheet is provided in Figure 2.3.

Dynamic Models
DC-PI models where time delays and feedback loops occur are of a more complex nature. Since the value of variables depends on their own or others' values in the past, they can best be evaluated using continuous dynamic models. Multiperiod evaluation models with feedback (Stella, Dynamo) could be used for evaluating relationships of this type. An example of a DC-PI model for evaluating HRM hire/fire, promotion and training strategies for long-term optimisation is presented in Figure 2.4.

Event-Flow models
Even more complex DC-PI models are those where the design choice changes the characteristics of events, and queuing of events influences the performance. The relationships between DCs and PIs can then only be evaluated by observing the operation of the real system or a model of it. Characteristics of the individual events can then have an impact on the behaviour of the system and are taken into account during evaluation. Event driven simulation tools (e.g. Smalltalk, Arena, Simmek, Exspect) could be used for evaluating relationships of this type. An example of a factory lay-out for simulating event flows through queues is depicted in Figure 2.5.

The use of various models
Only one or several DC-PI relationships can be built into one relationship model, the latter being preferable. This makes programming and execution more efficient and ensures that the characteristics of the same production systems are the basis of many evaluations. It is not

Human Resource Management modelframe conditions				
average demand on skill	1150			
demand on skill	2250			
demand on skill	3600			
demand on skill	40			
demand total	1000			
experimental input demand variation as fraction of average		0.50		
cost of subcontracting (relative to own hours)		1.2		
cost of training (expressed in hours work)		0.02		
capacity lost due to cross-training		0.05		
design choices				
level of own capacity (0 to 1)	0.90			
fraction generalist capacity (0 to 1)	0.15			
performance				
cost of specialist capacity	765			
cost of basic generalist capacity	142			
cost of subcontracting	168			
cost of training generalists for skill mismatch	1			
total cost	1076			
per cent increase	7.6%			
capacity lost due to generalist inefficiency	6.8			
capacity lost due to lack of work	56.3			
capacity lost due to specialist skill mismatch	64.2			

Fig. 2.3 Example appearance of an Analytic model in a spreadsheet (cf. chapter 33)

possible, however, to achieve the implementation of all relationships in a single simulation model, especially when the time frame of the model is important. It is not useful to build into one model DC-PI relationships that take effect on different time scales. Therefore, the short-term HRM model, which operates in a time frame of weeks or

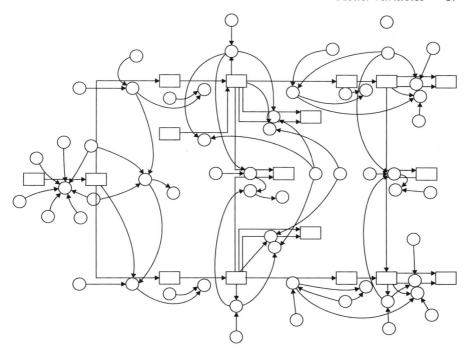

Fig. 2.4 Example of a Dynamic model

months, is kept separate from the long-term HRM model, which operates in a time frame of years to decades.

2.5 MODEL VARIABLES

The basis of the DC-PI modelling approach is that a relationship model evaluates the consequences of "design choices" on certain "performance indicators". During execution using "experimental variables", the model takes into account certain "frame conditions". Using the model to find the best design alternative, the design choices are kept fixed, while the experimental inputs are changed. After that, the model is executed with various sets of experimental input, e.g. various demand and supply patterns; different design choices are made and the experiments repeated. In this way, the best design alternative is found. Figure 2.6 depicts the relationship between various types of variables and the DC-PI model.

18   A design-oriented approach

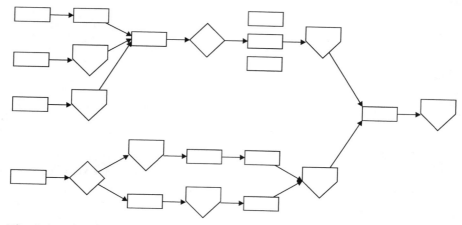

Fig. 2.5   Example of an Event-Flow and queuing model

Design Choices
Design choices (DCs) are the model variables whose effect we want to assess using the model. Usually the DCs are changed systematically between model runs (according to the experiment design), so results can be compared.

Performance Indicators
Performance indicators (PIs) are the relevant results of the model. They are used by the experimenter to assess the success of a particular design choice.

Frame Conditions
Frame conditions (FCs) are the parameters and assumptions of the model. They determine the strength of effects in the model. FCs are usually chosen for the duration of the series of experiments. A sensitivity analysis of a model for its FCs is desirable at the start of a series of experiments, though, to assess the danger of getting wrong conclusions due to small miscalculations in the parameters and assumptions.

Experimental Variables
Experimental variables (EVs) represent the uncertainty from the world outside the model. During execution of the model, several data sets are usually applied to the experimental variables to represent various external conditions.

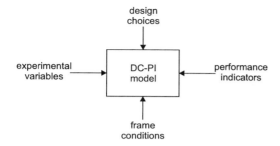

Fig. 2.6 Various types of variables and the DC-PI model.

2.6 SUMMARY AND CONCLUSION

The DC-PI modelling approach provides the tools for a multidisciplinary assessment of the effects of choices made in the redesign of a production system. By connecting production system design choices with performance indicators, it provides a clear insight into the effects of changing the production system design on various outcomes.

Through DC-PI models, it is possible to capture theory from various fields in a practical manner. It can easily be shared to complement the knowledge of various people, and can easily be used in concrete settings by practitioners in the field.

The DC-PI modelling framework provides the opportunity for combining detail with overview. It does so through its multilayered structure, where detailed, monodisciplinary relationship models fit into the framework of the multidisciplinary conceptual reference model.

The DC-PI modelling framework is of an open nature. An infinite range of modelling and simulation tools can be employed, as long as they clearly relate Design Choices to Performance Indicators that occur in the conceptual reference model and are explicit about Frame conditions and Experimental Variables.

The key value of the DC-PI modelling approach is that performance indicators from various fields of study can be related to a particular design choice. This allows an interdisciplinary, multicriteria evaluation of alternative designs, resulting in a better overall factory.

3

Customer driven manufacturing in a macro-economic perspective

3.1 INTRODUCTION

In the early years of the 90s the industrialised world was going through an economic recession. In particular the traditionally well-established branches were suffering from a severe downturn. It was not long after the economists and social scientists had been talking about post-industrial society as the dawning era. Manufacturing was expected to be relocated to less developed low-wage regions, leaving the service sector as the major source of livelihood in the western world.

As proof of such a trend, economic statistics showed that industry was leaving western societies. Additionally, one branch after another lost its lead in Europe and the US. Politicians, leading economists and even industrialists daydreamed about a new society without manufacturing, and invested in banking, insurance and stock markets in general. Acquisitions and divestments became the major means for development. Far too often there was no sincere aim to hunt for long-term returns on the products of manufacturing companies but instead just to make short-term profits through their stocks. The only manufacturing paradigm considered acceptable was the fully automated one. The early years of the 90s have taught us a bitter lesson. Areas, nations and companies live on manufacturing - not services. As Dertouzos writes sharply[1]:

": ... reject the much touted notion of a post-industrial society, where services are the exclusive answer to generating wealth. Without manufacturing there can be no wealth, and no services!"

Manufacturing industry is not homogeneous in its wealth-generating capability. The big multinationals have had a tendency to move towards the mass production paradigm with high market shares. In the 90s the "average" company has abandoned all diversification and streamlined its product mix to catalogue items. There is nothing wrong with cautiousness in diversification, but what about just the standard products? Between 1965 and 1991 major industrial players France, Japan, Germany and the US withdrew from the industrial investment goods sector to a relative degree of around 50%[2]. And it is the industrial investment goods sector that is the one with customer specified/modified products, if any. Is this trend also linked to the downturn in western economies?

When studying the competitiveness of nations, M. Porter noted that[3] a nation's competitiveness depended on the capacity of its industry to innovate and upgrade. This requirement for innovation and upgrading concerns products and processes. These ideas have led us to reassess customer driven manufacturing, giving it the role of an indicator for long-term economic success. Customer driven manufacturing should be seen as generative for all industrial activities because it searches for innovation in the case of every single customer order. In customer driven, make-to-order production the innovation capability is under continuous test, thus providing a yardstick and learning arena for repetitive production as well.

3.2 MASS PRODUCTION AND MASS CUSTOMISATION V. GENUINE CUSTOMER ORIENTATION?

The globalisation of markets seems to be reducing culturally based peculiarities in existing consumer products (e.g. even French cars look, feel and act international these days). This results in global marketing directed towards the product image to be created in the mind of a world citizen (potential customer). This trend may lead to a reduction in product variety, with the resultant colour of life being a simple grey.

However, it is questionable whether future society will be based on the production and consumption of harmonised products with less and less novel value. Rather, it is postulated in this book that as a counterforce to mass production there is a place for "genuine" customer orientation competing against the global standard product market share chasers. Even in consumer markets there will be end-customers who dislike this grey society and demand something better - variety. This group of *variety chasers* will generate a different class of factories of the future. It is challenging, due to high customer orientation and low global market shares (cf. counter reasoning behind the findings

in the PIMS database) plus low volume production owing to continuous customisation. How such factories of the future could be made productive and profitable is the subject of the following chapters.

Another justification for the survival of customer driven manufacturing as the fortress of the left over niches outside the mass customisation paradigm is related to quality. Superior quality can compensate for many of the demerits of a low market share[4] (figure 3.1). Such niches exist, at least in the capital goods sector.

Mass production and customisation factories are highly applicable arenas for cost-based management. Therefore, the cost-oriented *lean production principles* are applicable as well. Instead, the basic rationale of "genuine" customer-oriented factories is the creation of unique value to the customer. For this reason, later in this chapter the phrase **value-adding** *customer driven* **manufacturing** is systematically used as a generic term for this class of factories. Value-adding customer driven manufacturing is neither overweight, nor slow, nor does it carry the illness of anorexia nervosa like a lemon squeezed to the very last drop.

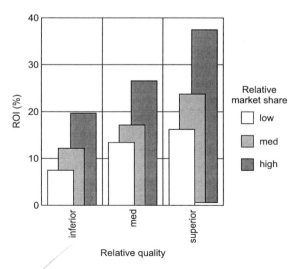

Fig 3.1 The impact of quality on ROI in different market share segments

3.3 BEYOND THE HYPE

Robert Eccles and his co-authors in their book "Beyond the Hype"[5] wholly express the weariness of academics and practitioners with the "flavour of the month" phenomenon. Themes such as just-in-time, theory of constraints, time-based competition, lean production, activity-based management, logistics, kaizen, computer integrated manufacturing, re-engineering, business process management, concurrent engineering, visual management, benchmarking, etc. come and go as the next consultancy enters a corporation while the previous one is still collecting its bill. The flow of these "isms" is accelerating and the managements of companies are asking themselves what each of them is about.

Currently, the only sensible solution is perhaps to make a break and try to concentrate on the essence, as Eccles et al. are proposing. The future of a factory certainly does not lie in the imitation of any of the hypes. The future of any factory lies in the *operations that add more value than cost*. Even this trivial observation is not sufficient. For survival, a *continuous flow of orders* naturally has to be won.

Product characteristics, *price* performance, *time* (delivery time and reliability) and *quality* (operations and product) are the criteria according to which the customer positions us as the best of all (we win the order) or not the best. The Olympics is a gentle and easy race compared to global business. In every race in the Olympics at least the best three (in boxing, four) win a medal. In global business only one wins, all the others (perhaps even the order winner!) lose.

The message common to all the hypes mentioned above is to concentrate on the *real processes*, "Go for Gemba!" as one of the isms, namely kaizen[6] (continuous improvement), explicitly proposes. When developing a company's real processes, the current hard times make it possible - and wise - to carry out only those development projects that involve low investment costs, low risk and have good cost-saving potential.

In the discipline of industrial management there are far more beliefs than scientifically solid facts. Among the current hype, however, there seem to be at least two themes that have broad factual evidence of validity. These themes are *quality and speed*. The underlying reasoning for quality and speed issues is based on their concentration on the real processes. Additionally, there seems to be more and more evidence that the driving thrust for quality and speed stems directly or indirectly from an atmosphere of good leadership and empowerment. We should create a strong

> ***hypothesis*** *that the survival of customer driven manufacturing is based on the foundation of improvements in leadership, quality and speed.*

The scientific evidence for quality and speed is discussed below. The crucial issue of leadership is also considered in the subsequent lines.

3.3.1 Quality

The PIMS database suggests that quality and profitability correlate with statistical significance. Through greater elaboration it can be observed that in any segment of market share, good relative quality results in better profitability (figure 3.1). Thus, it can be postulated that good quality can compensate for the deficiency of low market shares. This is an important message for customer driven manufacturing.

The IMVP (International Motor Vehicle Programme) data also indicated the correlation between quality and physical productivity[7]. However, this data is limited to automobile manufacturing.

A point of warning is also worth mentioning when talking about quality. Sometimes the quality movement has the symptoms of a religion rather than a business management tool. Just as any mass religion sometimes has traps, the quality movement may suffer from:

- form over spirit. Merely by following the rituals of the movement, nothing tangible is actually gained. For example, there are too many quality system certificates that have a negligible impact on actual quality;
- hypocrisy. People speak of quality but in practice act differently;
- blind obedience. Shutting one's eyes to other approaches that could benefit the health of the company.[8]

The above words are just warnings about the misuse of the quality movement. An active, personnel-driven quality movement has been successful, productive and profitable in a tremendous number of cases.

3.3.2 Speed

There is a common misunderstanding among practitioners that fast and flexible manufacturing systems are easily controlled but insecure in economic justification. Such misunderstandings have, however, been proved false in many individual companies in the past, and more recently via more general scientific justification, too. For example, the statistics collected in the PMS database show clear evidence of the harmful effects of inventories on ROI, i.e. the demerits of being slow.

Lehtonen[9] has provided further proof of the relationship between responsiveness and profitability. In empirical studies among mechanical engineering and foundry companies Lehtonen observed a clear positive correlation between ROI and the time-based competitiveness of the sample companies. This relationship was linear in the logarithmic scale, indicating that if a fast factory gets even faster it wins more in ROI than another, slow factory gaining an equal improvement in speed.

The correlation of physical productivity and speed has a strong correlation in some branches of industry. Holmström[10] collected UN industrial statistics and confirmed that at least in automobiles, telecommunications and office and computer industries "time is money".

3.3.3 Misinterpretation of Lean Production

One of the hypes still vital in its penetration is lean production, generated as the result of the MIT IMVP programme[11]. The lean production hype has been a major success worldwide. In industrial organisations the deep recession together with the lean production hype has been a devastating weapon. When companies lay off, say, 30% of their personnel, we should ask ourselves whether this trend has not gone too far. The interpretation of lean seems to be - at least in the extreme cases - the minimisation of all costs! This, naturally, was not the original scope and intention of lean production. A *distinctive feature of the original lean production lies in the people aspect,* just as Dertouzos later indicated in his list of "global best practices". The lean production hype must not be misused as the cost minimisation paradigm. If only costs are minimised, the company might face the side effect of minimum value added!

3.3.4 The New Breed of Leaders

The essential difference between value-adding customer driven manufacturing and the current stereotype lies in the corporate leadership culture.

Value-adding customer driven manufacturing is lightweight in the sense that it is resource saving. This means that separate administrative, directive and development resources are scarce. The management culture of the factory of the future is completely different from the average state-of-the-art factory. The leadership role of every manager should change. The role of the foreman (in its new meaning, scope and volume) as a military sergeant should be replaced by the new role of a coach and trainee. Rule obedience and cost minimisation at any cost is not part of the value-adding game.

When considering the problems of improving productivity in the 90s, Peter Drucker[12] emphasises the error of replacing humans by automation in all but simple tasks. In value-adding customer driven manufacturing every worker from the machine operator to the salesman, designer and top manager possesses the integrated roles of factory, knowledge and service work. In this integrated role automation cannot replace man but only support him, thereby improving responsiveness, productivity and quality. The concept of value-adding customer driven manufacturing must not be just an empty phrase with mere technical implications. Its major characteristics are:

supervision	learning, knowledge, skill
management targets	attention, innovation
information	problem-solving capability
control	indirect (values, support)
objectives	evolution

The basic idea is to consider value-adding customer driven manufacturing as a prototype workshop, engineering office or artistic studio rather than a cloning machine (mass production or mass customisation plant). A cloning machine is a factory which, by pushing a button, produces a product or batch of products designed and planned beforehand. A studio is a factory in which innovation is required for every product. This means innovation in all aspects: design, production and operations. Unfortunately, the state of the art in personnel administration is dominated by knowledge of and experience with the copier analogy.

The idea of a new leadership culture is strongly supported by the research groups of the "Made in X" projects[13], where X stands for the USA, Japan, France and Sweden. These research groups came to a unanimous conclusion about the most competitive industrial best practices. These practices are the cornerstones of value adding customer driven manufacturing of the future. These best practices are the following:

1. A workforce that is broadly, well- and continuously educated, as well as properly appreciated and rewarded
2. Cooperation within companies, with suppliers and even with competitors
3. Mastering manufacturing technologies, especially in terms of new processes
4. Living in the world economy by caring for the mores and interests of all people and shopping for the best technologies and suppliers, wherever they may happen to be.

Dertouzos emphasises the central message behind the best practices above by stating that

"... the dominant focus of this new world of manufacturing is on *human beings*, with a secondary, yet important focus on *technology as a support* of the primary focus".

This reorientation of focus onto human beings when chasing competitiveness is most complicated: it is easily uttered, widely understood as lip service, but extremely seldom properly exercised. The redistribution of power in organisational hierarchies is a revolution that the current management consciously or unconsciously often is resisting - for understandable but unacceptable reasons. The motivation for this new regime of corporate culture is not to play humanistic, but to get competitive. Mutual trust, care and commitment is not created by superficial pretension, it is an order in heart and soul.

3.4 THE ROAD TO IMPROVEMENTS

The fundamental procedure for the development of value-adding customer driven manufacturing is quality, time or total productivity driven (depending on the reader's preferences about the hypes). The emphasis should be on proaction, prevention, uncompromising customer focus and speed in a development procedure driven as a movement covering the whole organisation. The commitment to development should be top down. However, the *development procedure is and should be bottom up.* There is no other direction for improving the real processes. No middle manager or foreman knows how the customer is served in practice, how a purchased item is directed in the supply chain or how a piece of material is given added value. If the generals, colonels, lieutenants and sergeants hesitate to distribute power and responsibility in the development process, it is doomed to fail, and so it is better not to start such a project. The Japanese were perhaps by chance lucky in their invention of Quality Circles, i.e. a parallel hybrid organisation of vertically and horizontally collected individuals equal in the development team[14]. As Japanese organisational hierarchies are based on seniority, there perhaps was a need for a spontaneous bottom-up driven force for developing the company among the competent, ambitious youngsters who, organisationally, are low down in the paternalistic hierarchies.

The necessary grease in the development process is education, but not the traditional classroom instruction about business trends and hypes. What is required is on-the-job training in identifying problems and opportunities for improvements at the atomistic level of real industrial processes. The fundamental question is the effectiveness of work (in particular, what work can be left undone?). In value-adding customer driven manufacturing the effectiveness issue is tied not only to cost minimisation. A completely new question needs to be raised: how to create new added value to the customer! Everyone has a "constitutional" right and responsibility to develop his own work to produce more value.

A typical shape of development action is *teamwork*. Some of the development objects are cross-functional business process re-engineering[15] tasks that need the involvement of horizontally and vertically broad teams. Gradually, even the organisational dimensions, in particular the vertical ones, should lose importance - and to some extent even disappear. The same trend will also unify the colour of the collar. Only the business processes have relevance. The best shape for a customer driven business process is horizontal.

3.5 CONCLUSIONS

Value-adding customer driven manufacturing is a great challenge to industrial companies. Fundamentally, the keys towards better performance are

- appropriate strategic positioning
- increasing speed in supply chain operations
- improving quality in products and operations.

The necessary characteristics of the team members in value-adding customer driven manufacturing are

- multiple skills
- ability to perform simultaneous tasks (multiple team membership)
- ability to identify and solve improvement opportunities
- ability to work in a team.

Everyone in value-adding customer driven manufacturing should be able to calculate the effect of his/her work and development initiatives on the economic result of the company as a whole.

REFERENCES

1. Dertouzos M.: *Manufacturing and Competitiveness*, Opening Speech, in APMS '93, 5th IFIP W.G. 5.7 International Conference on Advances in Production Management Systems, 1993
2. Ahlström, K., CEO of Ahlström Corp., Presentation at the Annual Conference of the Institute of Industrial Automation, Helsinki Univ. Tech., 1992
3. Porter, M.: *The Competitive Advantage of Nations*, MacMillan Press, 1990
4. Buzell, R., B. Gale: *The PIMS (Profit Impact of Market Strategy) Principles, Linking Strategy to Performance*, 1987
5. Eccles, R., N. Nohria, J. Berkeley: *Beyond the Hype - Rediscovering the Essence of Management*, Harvard Business School Press, 1992
6. Imai, Masaako: *Kaizen*, Japan Management Association, 1988
7. Womack, J., D. Jones, D. Roos: *The Machine that Changed the World*, Rawson Associates, 1990

8. Buzell, R., B. Gale: *The PIMS (Profit Impact of Market Strategy) Principles, Linking Strategy to Performance*, 1987
9. Lehtonen, A.: *Time Based Competitiveness*, Ph.D. Thesis Manuscript, Helsinki Univ. of Tech., 1993
10. Holmström, J.: *Realizing the Productivity Potential of Expeditious Operations*, Helsinki Univ. of Tech., 1993
11. Womack, James, Daniel Jones, Daniel Roos: *The Machine that Changed the World*, Rawson Associates, 1990
12. Drucker, P.: *The New Productivity Challenge*, Harvard Business Review, Nov. Dec. 1991, pp. 69 - 79
13. Dertouzos M.: *Manufacturing and Competitiveness*, Opening Speech, in APMS '93 Conference, 5th IFIP W.G. 5.7 Conference on Advances in Production Management Systems, Athens, Sept. 1993
14. Lillrank, P. and N. Kano: *Quality Circles*, Gummerus, 1991
15. Robert C. Camp: *Benchmarking - the Search for Industry Best Practices that Lead to Superior Performance*, ASQC Quality Press, 1989

4

Why customer driven manufacturing

4.1 INTRODUCTION

Various changes have taken place over the last few years in nearly all areas related to manufacturing due to technical, macro-economic, micro-economic and political developments, needs and interests. Today's situation has been described as one in which "the market has taken the power over from industrial companies, which are now time-driven and customer-driven" (Gallois 1993). In response to this, the nature of production processes has shifted gradually over the past several years from mass production and large batch sizes to small batch sizes and customised production. The extreme case is a product that is manufactured only once. This type of production can be called "one-of-a-kind" or "customer-driven." Besides the unique customer wishes, requirements and customised product, the organisational and technical structures and processes needed to produce a one-of-a-kind product are often unique as well.

The characteristics and underlying principles associated with this type of production situation are described in the following section.

4.2 GENERAL DEVELOPMENT TRENDS IN MANUFACTURING

To identify the requirements that will determine the nature of future production systems as well as today's production system requirements, the general development trends that affect manufacturing and related systems need to be considered. There is a current trend to produce smaller batch sizes. The relatively simple products will gradually be replaced by increasingly complex products in the future. Also development lead times will become shorter.

With the change from regional/national markets to global markets, globally-oriented production facilities will replace the production facilities that now only cater to local markets. In this situation, companies will need to rethink their business strategies in terms of shifting from a competition orientation to a cooperation orientation. Since our "throw away" society is becoming more conscious of environmental issues, public pressure and market pressures are forcing companies to formulate global environmental responsibility policies with respect to their products and production processes. As a first consequence of this, companies are already being confronted with extremely high costs for waste disposal in some industrial nations.

Given this situation, future production systems are not likely to be designed to resolve isolated problems. Instead, a total concept will be emphasized in which a variety of environmental factors are taken into account.

4.3 GLOBAL CHALLENGES FOR FUTURE PRODUCTION SYSTEMS

The fundamental rule of growth is that when capital expands faster than the population, the standard of living rises. The contrary is true in Third World countries, however. There, economic prosperity is declining and the population growth rate is climbing. This increase in population hinders the growth of industrial capital, especially as an increased consumption of capital is necessary to maintain life-essential services. The cause for the dilemma of the Third World is an unequal geographical distribution of industrial growth. Economic growth remains concentrated in the already highly industrialised countries and a number of newly industrialised countries (NID). In addition, as the result of recent developments such as the disintegration of the major military blocs, industrial nations are being forced to undergo a radical change.

Only suitably globally-distributed prosperity can guarantee and/or improve prosperity for everyone, including the industrial nations. The flow of refugees from developing and overpopulated countries - as currently experienced in Europe and the USA - will not cease until these people see a chance of reasonable development and adequate prosperity in their own countries. The industrial nations need to ask themselves what the value is of a product if there are only a limited number of paying customers. This leads to the conclusion that to achieve a suitable globally-distributed prosperity within the restrictions of global political demands, industrial nations are being urged to consider a worldwide distribution of labour.

Despite the initiatives of industrialised countries in this context, it is inevitable that production resources will need to be reassessed in the future. Mass production and even batch production will continue to shift to the newly industrialised countries (NID), Eastern Europe, etc. because of factors such as cheaper labour and limited environmental regulations. This is a trend that will not be reversed.

A conscious and deliberate shaping of the global distribution of labour could support the introduction of environmental protective production methods and systems throughout the world, on the one hand, and could be instrumental in preventing national conflicts on the other hand. This means that future industrial production needs to be expanded through the use of conservational resources and environmental technologies. Only in this way can the earth's ecological capacity be preserved and the survival of future generations be permanently guaranteed.

In addition to these global correlations, the resulting business and technical challenges for industrial production are described in the following section.

4.4 ECONOMIC AND TECHNICAL CHALLENGES FOR FUTURE PRODUCTION SYSTEMS

The introduction of a new product on domestic as well as foreign markets is traditionally carried out according to the "Market Cycle Theory" (see Figure 4.1).

During the first stage, the innovation phase, the product is developed and produced for the domestic market. After sufficient marketing and production knowledge are built up, the product can then be introduced in foreign markets (export phase). In the third phase the production is improved and standardised. Very often it is then adopted by competitors or foreign producers who have the advantages of producing with less risk, lower labour costs and limited environmental regulations (imitation phase). The innovator's technological advantage at this point decreases gradually until he ultimately starts importing the product from foreign sources himself (import phase). Sometimes an additional stage, the repatriation phase, follows. This occurs when the country that initially designed the product can develop a new generation of the product that incorporates major improvements in the

36 *Why customer driven manufacturing*

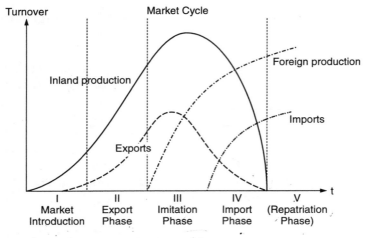

Fig. 4.1 Market Cycle (Dichtl 1991)

product or in the production technology. In such cases, domestic production is reestablished. This generally acknowledged theory emphasizes the close relationship between international competitiveness and innovative capacities.

In addition to this, Figure 4.2 shows the future shift of the R&D cost curves and the turnover in relation to the phases of the product life cycle. While the R&D costs increase, the market cycle and the turnover decrease. As mentioned previously, the highly industrialised countries will usually only be competitive in the early stages of production in the future. In addition, it is especially important for them to continue to develop so-called intelligent product or process innovations that

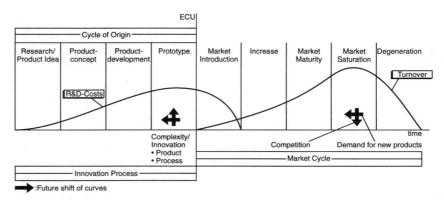

Fig. 4.2 Effect on the Product Life Cycle

cannot easily be imitated. These, however, generally depend on a high research and development input requiring extensive funding. As shown above, this pressure is increased by the fact that market cycles reflecting market penetration and customer demand are continuously becoming shorter.

The development, maintenance and enlargement of the prototyping capability to ensure a short "time-to-market" are becoming a crucial challenge for a modern Industrial Community. As a result, the initial product developer is likely to be the one to make the most profits. Consequently, in the future, particularly the rapid transformation of new ideas into prototypes (first-of-a-kind) and related manufacturing equipment will offer opportunities to achieve profits in the field of serial and mass production.

To meet these challenges, industrialised nations need to aggressively pursue advanced manufacturing technologies as well as take into account major organisational breakthroughs driven by advances in information and communication technologies. To specify the needs of future production systems, the industrial nations must define their future role in the context of an increased globalisation and distribution of labour and production facilities. The time has come to think about a new production paradigm for the industrialised countries.

4.5 CUSTOMER DRIVEN MANUFACTURING SYSTEMS

The statements up to now which have emphasized the "time-to-market" approach depict only one form of perfection or optimisation of the Mass Production Paradigm (MPP). The limits of the MPP will be reached by reducing lot sizes and increasing product variance. One illustration of this is the fact that there are so many possible configurations of the Daimler Benz 190 that only two identical cars are produced each year out of a total of 120,000 cars annually (500 per day) at the Bremen plant (Zeyfang and Hesse 1992).

Consequently, the necessity of a new orientation should be considered. Even if the majority of consumer goods will continue to be produced in large batches in the future, product individualisation will take place in more areas. The extreme case of infinite product variance is attained when each product is manufactured only once. This type of production is called "one-of-a-kind" or customer driven manufacturing (CDM).

With a lesser amount of reproductive manufacturing, i.e., product reproduction based upon existing product and process documentation, the direct application and/or operational realisation of a uniquely developed and manufactured product (a one-of-a-kind product) will be achieved.

With respect to capital investment goods, the constantly growing customer requirements for technically advanced capabilities and ease-of-use are leading to increased product complexity and uniqueness. The same tendencies are apparent in the area of luxury goods due to the increasing requirements for distinctive functions.

Because current "Rapid Prototyping Technologies" (RPTs) such as Stereolithography (SLA), Selective Laser Sintering (SLS) and Laminate Object Manufacturing (LOM) are limited in the range of materials, they are not suitable for building a one-of-a-kind product (a "final prototype") which can be used. The following statements focus mainly on the production of capital investment goods such as normally found in the manufacturing of machinery, plant design, shipbuilding, air and aerospace technology, etc. Characteristics intrinsic to this type of CDM business are:

- right-the-first-time design and manufacturing;
- customer intervention in all product life cycle phases;
- construction site manufacturing;
- complex product structures (process to product);
- universal and flexible production resources;
- international partnership (decentralised production facilities);
- decentralised engineering and information management;
- life cycle oriented engineering and information management;
- a high degree of high-grade concurrent engineering;
- high social and technological worker competence.

These characteristics are easily recognised in industrial practice. Based upon these secondary characteristics and reflecting the general development trends of the international markets, general requirements for the "prototyping capability" or "one-of-a-kind capability" of production systems can be identified.

To be competitive globally, prototyping capability means the ability to offer and manufacture incomparably unique sophisticated products based on continuously changing customer demands. This will require

- Information management for continous customer orientation
- Information management for life-cycle engineering
- Supporting structures for creative engineering processes
- Communication and information management for decentralized engineering
- Concepts, strategies and information management for handling uncertainties
- Systematic reuse of data, knowledge and experience during product definition, production planning and manufacturing
- Cooperation instead of competition (internal and external)
- Business integration
- Communication support for decentralized global production
- Dynamic order- and product specific factory layout
- Strategic performance assessment for company specific production system development
- Application of rapid Prototyping strategies and technologies for parts, units and products
- Continous incremental innovations

Fig. 4.3 Required Prototyping Capabilities of Future Production Systems

a focus on the quality and productivity of intellectual workers; less automation but higher levels of creativity will be demanded.

Although various industries are already faced with the requirements of customer driven manufacturing (CDM) there is only limited knowledge about the principles and related best practices. Although a systematic reappraisal for a new production paradigm is beyond the scope of this chapter, some prototyping or one-of-a-kind capabilities are highlighted in the following sub-sections.

4.5.1 "Do it right the first time"

Both in mass and serial production, a decrease in the learning curve can be achieved by incremental investments in the automation level of production systems. The greater the product quantity, the lower the price of the product. The lower the price, the greater the need for low production costs. The lower the budgeted costs, the greater the need for a high automation level.

Depending on the capabilities and tradition of industrialised nations in production management, there are different approaches (I, II, III) for managing the correlation between cost per component and the number of components as characterised by the different learning curves (Figure 4.4). Nevertheless, all of the approaches are incremental based upon the idea of optimising the production process and reducing the costs per component through optimisation of reproduction strategies.

40   Why customer driven manufacturing

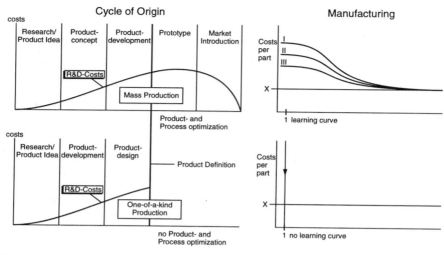

Fig. 4.4 Mass Production versus One-of-A-Kind Production

In the case of one-of-a-kind, however, there is no opportunity for optimising the production process through reproduction. The ultimate challenge then becomes "do it right the first time."

4.5.2 Flexibility: from technical integration to human cooperation

With decreasing lot sizes and increasing customer intervention the power of production systems is shifting from technical integration to human cooperation. Due to the uniqueness and complexity of customer driven products, the production planning, production resources and human resources need to be highly flexible. The one-task/one-employee principle in Adam Smith's pin factory is not adequate here. Adam Smith's doctrine is based upon standardisation and specialisation, but the CDM business implies nonstandard and fuzzy situations. This requires capabilities that are so complex and flexible that a single employee could normally not cope with this. This means that highly qualified personnel, group work and the configuration of stable work teams are essential requirements (Hirsch et al. 1992). The cultivation and employment of human experience become an essential subject of modern production systems when the competence of the worker is essential and needs to be maintained through suitable decision and responsibility structures. One main requirement is that this experience must be acquired and maintained during the daily work ("Learning by Earning").

4.5.3 Reuse of experience, knowledge and product data

If there is no possibility for testing and optimising the product and the production process, the systematic reuse of knowledge and/or product data plays a predominant role, in addition to experience, in reducing the technical and economic risks. By increasing the use of standard, purchased and previously used components, the level of uncertainty about a product can be reduced dramatically. A reuse-oriented or experience-oriented classification of design, planning and manufacturing activities to distinguish between previously solved, known and unknown steps is required at the beginning of the order processing or, even better, during the bid preparation.

Appropriate strategies for reuse-oriented product modelling or product configuration are required as well as strategies to evaluate and reduce the variety of the products. Approaches to reduce unnecessary variants are important, but approaches to avoid the requested uniqueness of a product are dangerous in an economy that is customer-driven.

4.5.4 Inter-organisational and intra-organisational cooperation

The requirement for reducing time-to-market as described above has led to efforts to identify product development tasks that can be performed simultaneously and to extend the just-in-time concept to include the flow of information within product development processes. Due to the decrease in production depth, this trend is not only confined to internal processes but also covers the exchange of external information between manufacturers and/or customers and suppliers. The terms "Concurrent Engineering" and "Simultaneous Engineering" have been used in recent years to describe this phenomenon.

The resulting degree of specialisation and integration between designing, planning, and producing a product has, therefore, not only become increasingly important on an internal level but has also become an essential part of the cooperation between manufacturer and supplier. In some sense, the supplier is becoming the (manufacturer's) external special department.

A special requirement for CDM in this context is that usually, even as early as the bid preparation phase, a network of independent, cooperating partners must be established. To ensure the necessary level of cooperation, appropriate information systems, communication technology and organisational structures are required. Internal and external data exchange as well as the functional inter-dependencies between the various business activities, therefore, need to be coordinated and integrated to enable the enterprises involved to deal effectively with a joint project in an integrated manner.

4.5.5 Customer orientation and integration

As mentioned previously, this contribution mainly focuses on the production of capital investment goods. In terms of customer orientation and integration, a one-of-a-kind or prototyping capability means the ability to offer and manufacture customer-specific (i.e., unique!), sophisticated products based upon requirements that are subject to continuous change.

In customer driven manufacturing, the customer's influence on the product can range from the definition of some delivery-related product specifications in advanced phases in the product life cycle (i.e., packaging, transportation) to a modification of the ultimate functions of the product in the very early product life cycle phases (i.e., customer-related product specifications).

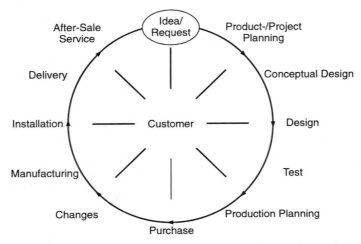

Fig. 4.5 Possible Customer Intervention in the Product Life Cycle

Systematic approaches are needed to derive appropriate organisational structures and production systems that can control customer orientation. Concurrent Engineering principles can be applied, integrating the customer as a full partner in all relevant phases of the product life cycle.

4.6 SUMMARY

Basic characteristics and principles of a possible new production paradigm called customer driven manufacturing (CDM) have been described in this chapter. Instead of avoiding customer-specific, unique solutions, the future production systems will need to have control over all aspects of the uniqueness of a product. The customer's wishes and requirements are the starting point for all these activities. A relationship of trust must be developed, focusing on a smooth-running cooperation between customer and manufacturer. Consequently, CDM requires a declaration from the companies for solving individual and unique challenges instead of delivering standard or quasi-standard solutions.

Although various industries are already faced with the requirements of CDM, there is only limited knowledge about the principles and related best practices. Ongoing production research activities must be strengthened to fill this gap in the future. Aiming at enforcing the prototyping and customising capabilities of future production systems, the appropriate engineering and production management systems, product information systems and related organisational solutions still need to be developed.

REFERENCES

1. Gallois, P.-M.: The new industrial challenge: Towards a "Lean Production" in *Advances in Production Management Systems*, I.A. Pappas and I.P. Tatsiopulos (Editors), Elsevier
2. Dichtl, E.: Der Weg zum Käufer, *Beck-Wirtschaftsberater im dtv*, 2. überarbeitete Auflage, Verlag C.H. Beck, München 1991
3. Zeyfang, D. and Hesse, R.: New organizational Structures for Automobile Assembly in *"One-of-a-Kind" Production*, Hirsch, B.E. and Thoben, K.-D. (Editors), New Approaches, Elsevier Science

4. Hirsch, B.E., Hamacher, B. and Thoben, K.-D., *Human aspects in production management*, Computers in Industry 19 (1992) p.65-77, Elsevier
5. Lische, C., *Abstimmung und Koordinierung von unternehmensübergreifenden Prozessen der Produktentwicklung*, Dissertationmanuskript, Universität Bremen, Fachbereich Produktionstechnik, 1992
6. Hirsch, B.E. (Hrsg.), *CIM in der Unikatfertigung*, Springer Verlag, Berlin, Heidelberg, New York, u.a. 1992

5

Describing production situations

5.1 INTRODUCTION

This chapter establishes a framework for describing customer driven manufacturing. It introduces manufacturing concepts such as engineering, production, products, resources, activities, stages, information, decisions, etc., which are the foundation for the following chapters. A detailed description of these concepts will be given in section 5.3.

Three of the fundamental and interdependent activities in a manufacturing company are engineering (or product design), production and manufacturing management (figure 5.1).

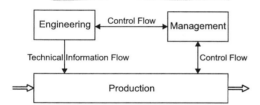

Fig. 5.1 Operational Processes in a Customer Driven Manufacturing Company.

Production is concerned with the flow of materials, the purpose of production being to transform raw materials into finished products. Engineering is concerned with the flow of technical information, usually represented as drawings or other documents, electronic or paper. The purpose of engineering is to provide technical specifications on what products to produce, and how to produce those products. Management is concerned, amongst other things, with the flow of operational information. This information is usually represented in the form of customer orders and work orders, or planning or status information. The purpose of management is to acquire customer orders and to deliver these within the agreed time, within the agreed budget and with the agreed quality. The means for

Describing production situations

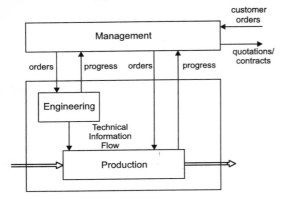

Fig. 5.2 Customer Driven Manufacturing consists of Engineering and Production

management are to plan, direct and monitor work orders for engineering and production. In addition, management has to acquire, develop and maintain resources.

Figure 5.2 represents the fact that customer driven engineering and production together constitute the manufacturing system that has to be managed. Thus, engineering is considered to be an operational task which produces documents and which is comparable to the production of physical products. Figure 5.2 also shows some information flows that are encountered in manufacturing. Information flows are of great importance for understanding the manufacturing system. For reasons of clarity, the interactions which exist between each of the information flows are not indicated in the figure. For example, there is no link shown between customer orders and the product design process, although such a link does exist, of course.

In the previous chapter of this book we have seen that management can be associated with strategic choices, structural alternatives and operational decisions. If we restrict ourselves for the moment to operational decisions (within the context of an organisational structure and strategy) the common way to describe management is depicted in Figure 5.3. This figure consists of:

- a controlled system: the manufacturing system;
- a controlling unit: human or non-human decision-makers, the decision-maker's organisation and the decision rules;

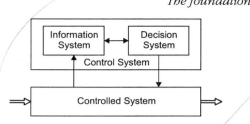

Fig. 5.3 Decision, Information and Control Model for a Manufacturing Company

- a management information system: the system which collects, processes, stores and disseminates the necessary information to support the controlling unit in making decisions.

Section 5.2 starts with a presentation of some basic concepts of manufacturing which can be related to a classical engineering-economic model of production (the so-called Walrasian model). The model is extended and the concepts are then combined into a general framework for describing Customer Driven Manufacturing in section 5.3.

5.2 THE FOUNDATION OF A FRAMEWORK FOR DESCRIPTION

A closer look at various production models in theory and practice reveals the interesting fact that they share some basic concepts. These concepts too can be related to the economic model of production, as developed originally by Walras. Let us therefore give a brief presentation of this model. A more comprehensive formulation and discussion of the model is given by Franksen in [3] and [4].

The Walrasian model depicts the process of transformation from *resources*[1] into finished *products*. The system considered as the production process or production function itself is represented by the network in the first quadrant. It can be considered from two viewpoints.

---

1 The term *production factor* is used synonymously with the term resource in the literature.

48 Describing production situations

Vertically, corresponding to each resource, it defines:

a *stage* or department as the parallel connection of products involved in the consumption of the single resource. This view can be called the resource view.

Horizontally, corresponding to each demanded product, it defines:

an *activity* or process as the series connection of resources involved in producing the single product. This view can be called the work-flow view.

The relationship, i.e. the *production function*, or ratio of transformation between the amount of a certain resource and the amount produced of a given product is known as a *technical coefficient*. In figure 5.4, a technical coefficient is symbolised by a cross surrounded by a circle.

As already stated, our understanding and the meaning of the two objects *resource* and *product* is quite open to interpretation. Take, for example, the three basic activities or functions of engineering, production and management. In engineering the product is, for instance, documents like drawings, and resources are human beings. The output from engineering then is technical information for production. In production the products are material and the resources are machines.

The meaning of the concepts can also be expressed by their functionality, namely: the function of the resources is to *support*, the function of the activities is to *execute*, and the function of the decisions is to *control*.

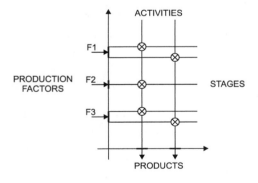

Fig. 5.4 The Walrasian Model

The above description of the Walrasian model is general and should be seen as a framework for the organisation of our concepts. The next section discusses the limitations and the necessary extensions to the model.

5.3 THREE EXTENSIONS TO THE MODEL

The next three sections give the extensions necessary to the Walrasian model in order to cope with real-life production situations. In particular, we will show how the product-oriented dimension in the Walrasian model can be extended into a "work-flow view of manufacturing", and how the resource-oriented dimension can be developed into a social and physical resource description — a "resource view of manufacturing". These two views provide us with the elementary concepts to describe manufacturing, i.e. the lower part of figures 5.2 and 5.3. Next, a third view needs to be developed, viz. the decision and control view, enabling us to describe the upper part of figures 5.2 and 5.3. Finally, the decision and control view and its impact on work flow and resources can be incorporated into one scheme or grid.

5.3.1 First Extension: A Work-Flow Description

Production is seldom timeless and performed in one stage or transformation only. Usually, a production system is built up of several manufacturing and assembly stages that take time. The original Walrasian production model does not consider the assembling of parts according to a bill of material or the sequence of manufacturing operations according to the routings as required for real production situations. That is, the production system in figure 5.4 should be extended over several levels by introducing an intermediate object, to be called *intermediate product*, and structuring it by the bill of material and routings.

A first step towards this extension of the product-oriented dimension of Walras can be achieved by including the Bill of Material (BOM) (see figure 5.5.a.).

The BOM graph is based on the *goes-into* or *where-used* relationships between the parts. For a given intermediate or other product the graph represents the number of components and subassemblies needed to produce one unit of the product.

50   *Describing production situations*

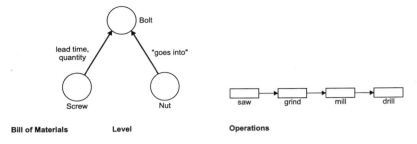

Fig. 5.5a   BOM graph   **Fig. 5.5b**   Routing or activity network

The second step towards extension of the product-oriented dimension of Walras can be achieved by including the network of activities or routing (see figure 5.5.b.). The activity concept (also called task, operation or work order, depending on the context) is fundamental for production scheduling. The activities are structured by the routing or precedence network (process plan, operation network, etc.). This is the other fundamental data structure in production next to the bill of material.

In some production control information systems, notably in so-called MRP systems, the routing information is subordinate to the BOM information. In other words, a routing is defined here as a sequence of operations for the production of a specific intermediate product from its components. These components and the intermediate product have to be defined in the BOM system.

However, this way of connecting the routing and the BOM is by no means the "one best way". If the routing is extended to an activity network, then all BOM and other information about the assembly structure can be described in the activity network. On the other hand, if the BOM would be extended such that after each operation a new (intermediate) product would be defined in the BOM, the BOM graph would contain all the routing information normally encountered in activity networks. We can conclude, therefore, that it is possible to represent the intermediate products and the operations required in two ways. These can represent the same information (when taken to the extreme), but can also be used jointly in a meaningful way.

In order to avoid these details in the present discussion in this chapter, we shall depict activities and intermediate products in one graph. As emphasised above, this graph represents both operations and parts simultaneously. Accordingly, the routing graph and the bill-of-material graph can be derived from this graph. In the following we will refer to this part of the production structure as the *P-graph* (figure 5.6). In principle, it is possible to develop other symbols in this graph in order to deal with several other issues, such as alternative materials or activities. However, in this introductory chapter further extension would defeat the object.

In practice, a graph such as figure 5.6 will seldom be encountered because of the need to distinguish between materials and activities. Therefore, the graph should be interpreted slightly metaphorically. However, the P-graph is a convenient concept for the joint concept of activities and bills of material.

The P-graph as described here represents the primary flow of activities and materials in the production part of the manufacturing process. In fact, it defines the work to be done. In this way, the product-oriented view can be generalised to a *work-flow-oriented* view.

The purpose of the engineering process is to establish the P-graph as defined here. Indeed, the engineering process itself can also be regarded as a production process, but with a different primary flow. In this case a similar graph may be used to describe the work flow of the engineering process. Note, incidentally, that the same also holds true for the management process.

In customer order driven production the P-graph will be developed gradually. Actually, production may start before the P-graph is completely defined. The amount of detail on the P-graph will also depend

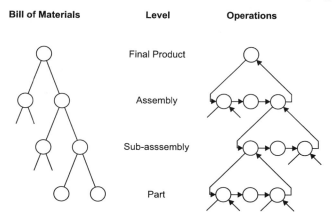

Fig. 5.6 P-graph

on the planning level we are working at. This is consistent with the fact that different activity networks and even different types of bill of material may exist in a company simultaneously.

The elements of the P-graph (activities or intermediate products) can be connected to the resources, as will be discussed later in this chapter. The P-graph in this way identifies the requirement for resources, this requirement being defined by a type of resource and by the amount required. The type of resource is called *capability* and the amount is often called *capacity*.

In an actual situation, demand for products is defined on a future timescale. It is possible to multiply the demand with the requirement figures of the P-graph, and then to time phase the resulting capacity requirements for resources and aggregate the requirements for the same capability in the same time interval. In essence, this process is repeated at different levels of detail in production planning and control. To summarise, we can look at the P-graph in four variants or interpretations:

A. Without timing
1. Bill of material - the graph identifies the physical components of the products. This is usually the result of product design.
2. Operations - the graph identifies all operations to be performed. This is the result of manufacturing engineering.

B. With timing
3. Operations - the graph is time phased and shows the operations required in different time periods, but without identifying the exact resource requirements and without fixing when the resources will be used. In MRP systems this is the result of material requirements planning.
4. Job orders - the graph identifies job orders. This includes time phasing and fixing resource requirements (both type and amount: capability and capacity). This is the result of scheduling and loading.

This extension is illustrated in figure 5.7.
These extensions of the product-oriented dimension of Walras into a work-flow description of the primary flow in both engineering and production is called the First Extension. For further details we refer to Part C, which is called Work Flow and Resources, and in particular Chapter 14 (Engineering), Chapter 15 (Production) and Chapter 16

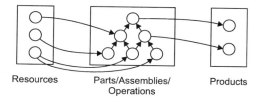

Fig. 5.7 First Extended Walrasian Model

(Bills of Material). The next section will present an extension of the resource dimension.

5.3.2 Second Extension: A Resource Description

The Walrasian model describes resources as a set of independent primitive objects. We need to be able to define how resources are combined and structured into groups, into departments and into factories.

A first step towards a more elaborated view on resources is, therefore, to allow the aggregation of resources, hierarchical or other, in a way which is analogous to the "bill of material", as shown in figure 5.8.a. This grouping of resources should allow us to describe the capabilities of certain resource groups and the available capacity within each group. However, the grouping of resources which can be represented by figure 5.8.a is rather fixed: it is not easy to represent the fact that humans or tools can be moved between cells or departments. In customer driven manufacturing it is often important to know which *types* of resource should be allocated to each other on a temporary or other basis in order to provide a certain capability. This leads to a second step towards extension of the resource-oriented dimension, illustrated in figure 5.8.b. This figure describes the structure of capabilities rather than the structure of resources. Together, the resources, capacities and capabilities are represented by the so-called *R-graph* by analogy to the P-graph (see figure 5.6.). As with the P-graph, the R-graph too can only be partly defined at the start of production. This is especially valid for human resources.

In principle, it is possible to devise other symbols in this graph in order to deal with several types of resource, e.g. machines, humans, tools, etc. However, in this introductory chapter greater preciseness

54   *Describing production situations*

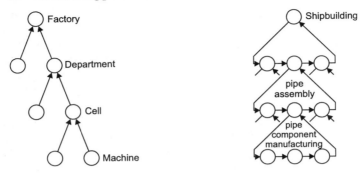

Fig. 5.8a   Resource hierarchy   **Fig. 5.8b**   Hierarchy of capabilities

would overshoot the mark. Therefore, the term R-graph is used slightly metaphorically. Part E of this book, which is devoted to information technology, will elaborate on more detailed data structures.

In an actual situation, the availability of resources will be mapped on a timescale and be balanced against resource requirements. The resource requirements come from customer orders and the derived job orders. The inclusion of the time dimension resource availability is established by knowing their basic capacity (units per time unit) and the working hours. These resources are allocated to the activities, i.e. an element of the R-graph is allocated to an element of the P-graph in

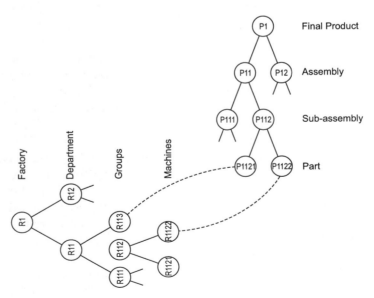

Fig. 5.9   Allocating activities to resources

a given time interval. This process is usually referred to as scheduling and loading (see figure 5.9). The dotted lines represent information on: 1) which tasks should be performed, 2) which capability is required, 3) which resources are involved and 4) when and how much of these resources are needed.

Summarising, these extensions of the resource-oriented dimension consider both group social structure and physical layout of the machines, like an FMS (Flexible Manufacturing System) for example. Extension to group/departmental physical structure, represented by the layout of machines with transport routes in between and transportation time assigned as lead time on the arcs, is well known. However, the extension to group social structure, i.e. what kind of relationships are to be represented, still needs to be discussed in detail. Chapter 17 of this book is devoted to a further discussion of resources with the emphasis on human resources.

We have called this extension of Walras to a resource description (social and physical) in both engineering and production the Second Extension.

5.3.3 Third Extension: A Decision Description

The Walrasian model is a model of the primary process of resources and products with a fixed production function in terms of technical coefficients. Two extensions of the model have been outlined to obtain a more realistic picture of work flow and resources. However, only engineering and manufacturing activities have been discussed in the preceding extensions. Information and decision processes, which are part of the management function, have not yet been considered. Fig. 5.10 adds information and decision-making to the extended Walrasian model.

The left part of figure 5.10 shows the Walrasian model (figure 5.4) with resources and work flows, and with a *time* dimension added. The right part of the figure shows in a grid how decision-making on work flows and resources can be distinguished but always require synchronisation and coordination. The columns of the grid represent the resources and the work flows of the Walrasian model as well as their connection (i.e. allocation of resources to products).

A column corresponds to the types of decisional activity. A line is characterised by a horizon and a period or length of time over which the decision extends (review period of the decision). The intersection

56 *Describing production situations*

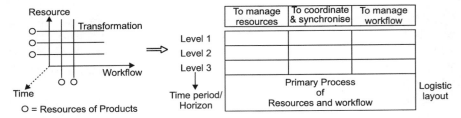

Fig. 5.10 From Walras to Levels and Aspects of Control

between a type of decision (column) and a decision level (line) defines the concept of Decision Centre. The structure of the decisions represents the time aspect of how decisions are related to each other, i.e. by time period and horizon. Thus, one decision sets the scope of the decisions at the next sublevel.

This way of representing decision structures is called the Grai Grid [8]. This grid, which is discussed in detail in chapter 18, provides a way of representing the multi-dimensional model on the left of figure 5.10 as a 2-dimensional grid, as shown on the right of figure 5.9. The Grai Grid represents mainly the controlling system. The controlled system, i.e. the primary process, is not usually depicted together with the grid. Instead, it is represented by the P-graph and R-graph of the Walrasian model, which were introduced above.

The dominant criterion in choosing decision levels should be the horizon over which decisions have their impact. It seems that the three levels of production control distinguished by Burbidge [5] (programming, ordering, dispatching) are also valid for the subject of this book, Customer Driven Manufacturing (where they are called aggregate production planning, factory coordination and production unit control).

The time dimension is included here in the grid in order to distinguish decision-making at several levels. At each level, decision-making consists of allocating resources and activities to each other on a specific horizon and with a specific level of detail. Let us therefore look more closely at the objects, resources and products, and see what type of constraints we can define to classify and subdivide them further. Resources are described in greater detail in Chapter 17.

Briefly, these objects and their constraints can be classified as follows.

Durable and non-durable
A durable resource, say, a machine, can be used over and over again as a service in a temporal sequence of jobs until it becomes waste and has to be thrown away. Economists therefore describe a machine as a *fund of services*. The constraints are that the capacity requirements of a machine per time unit for an activity must be less or equal to the available capacity of the machine.

A non-durable input, say, a material, is created as an output supply of a process or consumed as an input of a process. Analogous to a fund of services, economists describe materials as a *stock of supplies*. A non-durable resource constraint states that the consumption of material in any process must be less than the available stock of the material.

Tangible and non-tangible
The above two types of resource are tangible. Examples of non-tangible resources are ideas, skills, experience, procedures, knowledge and information. Non-tangible resources require representation by a data description in, say, a document. A non-tangible resource constraint states, for example, that information like a drawing is available for an activity before a certain due date.

Human and non-human
Human resources can be described as above as a durable resource (non-human). However, the main difference between human resources and non-human resources is that their performance with regard to physical power, skill and experience changes over time.

The constraint on the final products, i.e. the market of product demands, is that the price of the company's products will be less than or equal to the market price.

We shall call this extension of Walras to the decisional view of feedback and different levels of control of both engineering and production the Third Extension.

5.4 CONCLUSION

It has been argued that the three basic views in customer driven manufacturing are:

- work-flow view
- resource view
- decision view

and that organisation comprises all three views.

REFERENCES

1. Rolstadås, A.: *Structuring Production Planning Systems for Computer Applications.* APMS 1987, Elsevier, 1987.
2. Franksen, O.I.: Funds versus stocks: a note on constraint classification. *Production Planning and Control,* 1993, Vol. 4, No. 1, 84-87.
3. Franksen, O.I.: Mathematical Programming in Economics by Physical Analogies, *Simulation,* no. 4 (July 1969) and no. 2 (Aug 1969).
4. Franksen, O.I.: Introducing Diakoptical Simulation in Engineering Education. *The Matrix and Tensor Quarterly.* Sept. 1972. pp. 1-16.
5. Burbidge, J.L.: *Group Technology in the Engineering Industry* (Mechanical Engineering Publications Ltd, London, 1979).
6. Bertrand, J.W.M., Wortmann, J.C., and Wijngaard, J.: *Production Control - A structural and design-oriented approach* (Elsevier, Amsterdam, 1990).
7. Burbidge, J.L.: A Classification of Production System Variables. in Hubner (ed), *Production Management Systems: Strategies and Tools for Design.* Elsevier Science Publishers, IFIP 1984.
8. Doumeingts, G.: Méthode de conception des systèmes de Productique. *Thèse d'état en Automatique,* Laboratoire GRAI, Université de Bordeaux I, 1984.

6

A typology of customer driven manufacturing

6.1 INTRODUCTION

There are many types of customer driven manufacturing. All of these types may have common characteristics, as explained in the previous chapters, but they are not the same. There are big differences between a repair shop and an aircraft company or between ship building and truck manufacturing — and these differences are explained and highlighted in the present chapter. In this chapter, the nature of customer driven manufacturing is investigated by means of a typology, which is presented in Section 6.2

Section 6.3 proceeds with a characterisation of management problems in several types of the typology, focusing on production management. Section 6.4 discusses the consequences for information systems. The conclusions of this chapter are presented in Section 6.5.

6.2 THE TYPOLOGY

In order to develop a typology of production situations, we first have to introduce the notion of the Customer Order Decoupling Point (CODP). The CODP refers to the point in the material flow from where customer-order-driven activities take place. Stated differently, the activities upstream of the CODP are driven by planning activities based on forecasts rather than on firm customer orders. Also, logistics focuses on stocks. Information systems are based primarily on anonymous items, upstream of the CODP. Downstream of the CODP, logistics focuses more on time. Information systems are not only based on anonymous items, but also on customer orders.

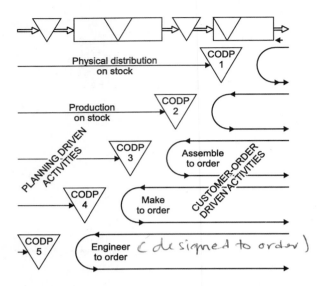

Fig 6.1 Typology of CODP

The concept of CODP is illustrated in figure 6.1, which is derived from Hoekstra and Romme [1989]. Different positions of the CODP give rise to quite different production situations. We shall not elaborate in this chapter on physical distribution, so the first position of the CODP is ignored in the remainder of this chapter.

Apart from the CODP, there is a second point which is important in understanding customer driven manufacturing. This point is illustrated in Fig. 6.2. It is concerned with the amount of investment made in developing products or production processes independently of the customer order.

A company is called **resource-oriented** (or **capability-oriented**) if it has invested substantially in resources (humans, machinery) but not in specific processes or products (independently of a particular customer order).

A company is called **product-oriented** if it has made substantial investments in product development independently of customer orders. Quite often, a product-oriented company has also invested in resources. However, we will still use the term "product-oriented" in such cases, because the resources are configured in a way which is determined by the product.

	Engineer-to-order	Make-to-order	Assemble-to-order	Make-to-stock
Product oriented	Packaging machines	Machine tools	Medical systems	Commodities
Workflow oriented	Printing	Fine paper	Service industries	Subcontractor of car-outlets
Resource oriented	Ship building	Repair shop	Construction company	Repetitive manual assembly subcontractor

Fig 6.2 Typology of production situations

A company is called **work-flow-oriented** (or **process-oriented**) if it has made substantial investments in production process development independently of the customer order. A production process consists of all manufacturing steps required to deliver a particular type of treatment. Quite often, a work-flow-oriented company has also invested in resources. Again, we will stick to the term "work-flow-oriented" in such cases, because resources will be configured in a way which is determined by the work flow.

Of course, there are also companies which are both product-oriented and work-flow-oriented. Thus, the three concepts used in this second dimension are not mutually exclusive. Such companies will exhibit a mixture of the characteristics of their "pure" counterparts, which will be discussed below.

The concepts of this second dimension are less obvious and less well-known. A few examples are therefore necessary.

Engineer-to-order
Consider an aircraft-manufacturing company. A company of this kind has invested billions of dollars in developing products without having a customer order. Nevertheless, companies manufacturing aircraft are usually willing and able to add some customer driven engineering to each individual aircraft sold to a customer. Therefore, an aircraft company is a **product-oriented** engineer-to-order company. The same holds for a company producing packaging machines (cf. Chapter 9).

The situation is different for an aerospace company producing satellites. A satellite is built completely to a customer's order. However, satellites are built by companies that are in a position to be able to build them right first time. A company of this kind is a **resource-oriented** engineer-to-order company. The same holds for most shipbuilding companies (cf. Chapter 12).

Finally, consider a printing company specialising in weeklies. Suppose that this company subcontracts all the work to other companies, but that it standardises and organises the flow of work. Thus, the company has invested in quality control and logistics, and in blanket contracts with other specialist, resource-oriented companies. It is able to subcontract not only the detailed layout and all the preparatory work, but also the printing and finishing and even the distribution of these weeklies. This company is a **work-flow-oriented** engineer-to-order company.

Make-to-order
A similar distinction can be drawn for make-to-order business. At first glance, a mechanical repair shop appears to be quite similar to a machine-tool manufacturer. However, the machine-tool manufacturer has usually developed his machine tools independently of any customer orders, whereas repair work is customer driven make-to-order business. Many specialist subcontractors, e.g. foundries, find themselves in the same position as repair shops. Thus, foundries and repair shops are **resource-oriented** make-to-order companies, whereas machine-tool manufacturers tend to be **product-oriented**.

The semi-process industry, such as paper manufacturing, is often **work-flow-oriented** business. As we will see in Chapter 11, the manufacture of fine paper can take the form of a work-flow-oriented make-to-order business. This type of industry has made high investments in the production process (and in plant equipment, of course). Another example of a work-flow-oriented make-to-order business is the manufacture of orthopaedic appliances, which have to be adapted to each individual patient. Here, too, work-flow-orientation goes together with investment in resources, but this time the key resource is craftsmanship.

Assemble-to-order
A similar explanation can be given to assemble-to-order business. There are products, e.g. medical systems, which require a considerable investment in product development by the supplier before customers can be approached. For this reason, a medical-systems manufacturing company is a **product-oriented** assemble-to-order company (cf. Chapter 10).

The nature of production management in different types of the typology 63

There are other products, such as houses and other real estate, where the supplier of these products has not made any investment in product development or work-flow development independently of customer orders. For this reason, construction companies are usually **resource-oriented** assemble-to-order companies - at least in Europe.

Again, **work-flow-oriented** assemble-to-order business is found, for instance, in semi-process industries, e.g. the production of fine chemicals or alloyed steel. Interestingly enough, work-flow-oriented assemble-to-order is often also found in service organisations such as restaurants or travel agencies.

This should be sufficient to explain the different types of customer driven business shown in figure 6.2. For the sake of completeness, we have also added some examples of make-to-stock companies with different orientations.

In the following sections of this chapter, the typology of figure 6.2 will be used to describe the differences in production management systems and information systems for several different types of Customer Driven Manufacturing situations.

6.3 THE NATURE OF PRODUCTION MANAGEMENT IN DIFFERENT TYPES OF THE TYPOLOGY

6.3.1 Introduction

In order to keep the discussion concise, we shall consider only four types, which four types will be discussed in the next four subsections. (N.B. Note that more detailed discussions will be found in the remainder of this book. In Part B of this book, four cases that almost match the cases of this introduction below are introduced):

a. product-oriented assemble-to-order business ("medical systems", see Chapter 10)
b. capability-oriented make-to-order business ("repair shop", see Chapter 12)
c. work-flow-oriented make-to-order business ("fine paper", see Chapter 11)
d. product-oriented engineer-to-order business ("packaging machines", see Chapter 9).

The discussion is summarised in figure 6.3. For the sake of completeness, this figure also shows a comparison with the product-oriented make-to-stock business ("commodities") in order to be able to highlight the peculiarities of customer driven manufacturing.

6.3.2 Production management problems in assemble-to-order

When a product-oriented company offers its products in an assemble-to-order mode, there is heavy competition in the **variety** of products offered to its customers. Usually, this means that product families are under continuous pressure for increasing variety and innovation. Thus, product innovation is a top-management concern. Slow innovation results in loss of market share, whereas unconstrained innovation leads to quality problems and high costs. This will become clear in the Medicom case discussed in Chapter 10.

In such an environment, operations management faces considerable uncertainty in forecasting the mix of customer orders that will be obtained. Therefore, it is very difficult to procure the right materials and to install the right capacities beforehand. Thus, the **assembly** of permanently varying customer orders is the most complex operation. This holds in particular in situations where material supply is not perfect, where customers change their requirements and where engineers change the specifications of the product — all at the same time.

For this reason, middle management in production focuses on the balance between material supply, capacity availability and customer-order entry. Ideas from flow production and Just-in-Time can be used to ensure capacity availability and short-term material availability. Long-term material supply can be supported by Master Production Scheduling in combination with intelligent product modelling (see Chapter 16).

Customer-order entry requires a number of steps which should be supported by information technology, such as checking for consistency of the customer's wishes, generating bills of material, manufacturing instructions and other engineering documents, calculating costs and quoting delivery dates based on the MPS.

6.3.3 Production management problems in make-to-order resource-oriented companies

The market situation for resource-oriented companies is usually more uncertain than for product-oriented companies. This is due to the fact that for resource-oriented companies there is no product available which can be marketed or forecast. The customers approach the company because of its reputation in production capability, price, responsiveness or service. Top management focuses on capacity and capability: the creation, improvement, maintenance and selling of capacity and capability. There is a strong need for a simple, rough capacity planning and monitoring system.

Orientation CODP Characteristic	Product Engineer-to-order	Workflow Make-to-order	Resource Make-to-order	Product Assemble-to-order	Product Make-to-stock
Top management's focus is on:	Customer order contracts	Process innovation	Capacities	Product Innovation	Marketing/distribution
Uncertainty of operations is concentrated in:	Product specifications	Volume of production	Work preparation	Mix of orders	Product Life-Cycles
Complexity of operations is concentrated in:	Engineering	Final production stages	Component manufacturing	Assembly	Physical distribution
Middle management's focus is on:	Project-management	Quality control	Subcontracting, Shop Floor Control	Master Production Scheduling and customer order contracts	Stock control
Information systems for P.M. are focussed on:	Support of Product Engineering	Progress control	Support of Manufacturing Engineering (work preparation)	Support of material supply and order entry	Support of forecasting and stock control
Nature of IS oriented towards:	Generative solutions	WFM	Reference solutions	Rules	Decision support

Fig 6.3 Characteristics of different types

The uncertainty of the operations is usually concentrated on the question of how much effort the production process for a particular job will take. Some companies avoid the risk caused by this uncertainty. These companies do not sell a product or service to their customers, they sell the capacity itself. In Bertrand et al. [1990], a company of this kind is referred to as a **capacity selling** production situation.

Typically, the complexity of the resource-oriented make-to-order business is concentrated in **component manufacturing**. Often, these production companies have limited capacity in order to limit their overheads. Consequently, there is a tendency to accept more work than available capacity allows and to subcontract part of the work. Therefore, middle management in production focuses on shop floor control and subcontracting. Issues such as lead-time control and efficiency are quite important. In many cases, some form of project management and project control is implemented in order to provide service to the customer. However, sophisticated planning and scheduling tools are seldom a success because of the many uncertainties: reliable and skilled personnel which are used to teamwork without avoiding responsibilities is in many cases a better guarantee for success. In other words, selling capacity in a make-to-order business is linked to appropriate human resource management.

As mentioned above, much uncertainty on the shop floor is due to a lack of reliable engineering data about the operations of new orders. Therefore, information systems which support manufacturing engineering (work preparation) are most useful. Note that such systems are completely different from the material-oriented information systems known from make-to-stock production. We shall come back to this point in section 6.4.3 and in Chapter 27.

6.3.4 Production management problems in make-to-order work-flow-oriented business

As already stated, make-to-order business that is work-flow-oriented often also involves investment in resources. The fine-paper manufacturing case that will be discussed in Chapter 11 is an example.

Companies that have invested in manufacturing technology are always threatened by competitors who could employ newer technology and start price competition or quality competition. Therefore, top management is often concerned with the **innovation of manufacturing technology**. However, work-flow orientation involves much more than manufacturing technology. It may include marketing and sales processes, customer service activities, purchasing and many other business processes. Whenever such processes are critical to a company's success, they deserve top-management attention for innovation. Similarly, work-flow innovation deserves top-management attention in service organisations like hospitals and universities.

Investment in technology (and often also in resources) usually requires long lead times before becoming effective, and still longer lead times before being written off. These long lead times contrast with the short lead times accepted by the market. In many cases it is quite uncertain in the short term whether the right **volume of production** will be obtained.

Customer driven work-flow-oriented manufacturing is often **complex in the final operations**, where increasing product differentiation occurs and where **quality problems** become apparent. This is a major concern for middle management.

6.3.5 Production management problems in engineer-to-order product-oriented business.

When discussing engineer-to-order in this chapter, we are concentrating on engineer-to-order business with significant product development, independent of customer orders. Chapter 9 describes this type of business on the basis of a company producing packaging machines. In this type of business, the **product strategy** should be formulated explicitly and communicated throughout the company. The product strategy specifies the features or components of a product family on which it is possible to negotiate and the limits within which customer demands can be met. Even if a well-formulated product strategy exists, there is a risk. Each customer order carries the risk of requiring considerable product development in some area, while being sold as a minor variation on a standard product. Due to these risks, top management usually focuses on the individual contracts with customers.

This type of business faces a fundamental uncertainty about the content of certain customer specifications. Usually, these specifications cannot be investigated in depth during the contract negotiations. Therefore, there remains much uncertainty about the nature of these specifications until the product engineering work is finished. The cost, lead time and the quality of the final product are largely determined during the engineering phase. Thus, production management cannot be restricted to management of the operations after the product engineering phase. On the contrary, product engineering is often the most complex operation to be managed.

The focus on product engineering also has another important management consequence. It means that the nature of production planning activities is not primarily material-oriented (as in assemble-to-order) or capacity-oriented (as in make-to-order) but project-oriented (or work-flow-oriented). Project management should not be associated with formal planning techniques, but rather with mastering the complexity of the customer driven activities, especially in product engineering. This includes, for example, document control, change control, quality control risk assessment, reuse of experience and an optimum use of human capabilities. An important point here lies in the support of information systems in defining and maintaining standard **engineering processes** and standard solutions.

6.3.6 Preliminary conclusion

Summarising the above, a few general statements can be inferred. When moving from make-to-stock towards engineer-to-order, top-management's concern with individual orders increases. Furthermore, the focus of production management shifts from material-oriented via capacity-oriented towards work-flow-oriented. The nature of the uncertainty faced by operations management changes from well-quantified and repeatedly occurring uncertainty towards qualitative and unique types of risk. Consequently, decision support in terms of mathematical models loses its relevance and fast feedback from the engineering operations is much more important. In engineer-to-order business, production management is highly intertwined with customer driven engineering.

6.4 INFORMATION SYSTEMS FOR ONE-OF-A-KIND PRODUCTION

6.4.1 Introduction

Within the context of this chapter, a complete treatment of information systems supporting production management in Customer Driven Manufacturing is clearly impossible. A more complete treatment is given in Part D of this book, Chapter 27. In this chapter, some points that follow from the discussion in the previous section will be highlighted. These points are summarised in the lower part of figure 6.3.

Before discussing information systems in assemble-to-order, make-to-order and engineer-to-order in more detail, it is useful to characterise information systems for make-to-stock.

Typical information systems for make-to-stock situations are MRP systems in production and DRP systems in physical distribution. These systems measure the flow of products and the flow of orders in terms of **anonymous products**. These products and their inventories are assumed to exist independently of customer orders. The aim of MRP-I and DRP-I is to generate automatically planned work orders for the manufacture or transportation of anonymous products. This leads necessarily to the assumption that engineering data and logistics data are complete, consistent and correct.

In customer driven business (especially engineer-to-order), the highest aim of an information system is not the automatic generation of planned work orders, but the user-friendly support of engineering professionals. This will be argued below.

6.4.2 Information systems for assemble-to-order production

In assemble-to-order production, the goods flow consists of two parts. The part upstream of the customer order decoupling point (CODP) displays much similarity with make-to-stock production, and the information system resembles the make-to-stock information systems just described. The part downstream of the CODP is quite different and will be described first. We shall draw a distinction between transaction processing and decision support.

Transaction processing
With regard to transaction processing, an important requirement is that it should be possible to trace the goods flow **in terms of customer orders rather than in terms of anonymous product codes**. There are several reasons for this statement.

First of all, in assemble-to-order business the number of possible final products within one family is usually far too large to make anonymous codes for final products useful. Therefore, the goods flow should be measured and traced in terms of a customer order and not in terms of an anonymous code number (see Chapter 16).

Furthermore, a product code is a shorthand for a completely specified product, both in terms of the wishes of the customer and in terms of the supplier's technology. However, product innovation is usually so fast that the same set of customer's wishes can often be met by different technical means. This choice should not complicate the sales process. Thus, codes for customer's wishes should not be mixed up with a particular technical means. Anonymous codes in the sales contract are harmful in this case.

Decision support

As for decision support, two interacting and difficult decisions have to be taken in assemble-to-order. On the one hand, material supply has to be provided for — long before firm customer orders are available. On the other, customer-order entry should be supported, presumably in relation to the materials and capacities which are being supplied.

The first decision has been located in the Master Production Schedule by MRP literature (see Orlicky et al. [1973]). This approach states that the customer's choices can be clustered into orthogonal **features**, each of which has a number of **options**. The MPS for the options can be derived from the production plan for the family as a whole by using the so-called **planning BOM**. The components of the product family should base their gross requirements on the MPS for the options according to the planning bills of material. It has been known for a long time that the features-and-options technique suffers from many disadvantages, but no alternative was available. Recently, a more general approach has started to emerge in which bills of material, routings, instructions and other documents are **generated** rather than selected. This approach looks rather promising (see Hegge [1995], Van Veen [1991]) and is discussed in more detail in Chapters 16 and 20.

The other important decision, customer-order acceptance, should also be supported by appropriate software. First of all, the wishes of the customer should be checked for completeness and compatibility. Next, the resource requirements of this particular customer order should be generated and compared with the availability. This results in a proposed due date and price.

6.4.3 Information systems for make-to-order production

Let us briefly recall what section 6.3 revealed about capability-oriented make-to-order companies. These companies focus at top-management level on capacity. Production management concentrates on issues such as manufacturing engineering, shop floor control and subcontracting.

Rough planning is required to support negotiations with customers, and detailed planning is required for managing the capacity load and for subcontracting.

In all these cases, manufacturing engineering is essential in order to obtain information on the relevant capacity requirements. Therefore, the key element in information systems lies in the support of manufacturing engineering.

A reference base with manufacturing engineering data

Information systems supporting manufacturing engineering should provide tools for quick data entry at various levels. Consider first the level of rough planning. Networks with activities for projects which are in the negotiation phase should be created fast and should be easily modified. For capability-oriented companies it is unlikely that such networks will be generated automatically. Rather, there should be a database with former projects which can act as a starting-point for engineers to create a new project. It goes without saying that a database with reference networks should be maintained carefully in order to guarantee that this "reference base" contains the accumulated best experience in the company. It should prevent future projects from making the same mistakes as earlier ones. The same is true at the level of detailed planning. Again, information systems should act as toolkits in order to make manufacturing instructions and routings readily available.

Fast production of these documents enables the company to take subcontracting decisions at an early stage. Automatic generation of these documents will seldom be possible because the company is capability-oriented, but the existence of a reference base with applicable solutions and with intelligent software support becomes within reach of many companies.

6.4.4 Information systems for engineer-to-order production

Engineer-to-order production for a product-oriented company has been characterised as follows in section 6.4. Top management focuses on customer-order acceptance, operational management on customer-order-specific engineering activities. The reuse of knowledge and capitalisation of experience is a key-issue here. So, engineer-to-order largely relies on people, and human resource management is a most important aspect of management.

For a **product-oriented** company it can be expected that the set of well-defined engineering activities will be limited. Usually, a product family in an engineer-to-order business offers a number of functions in various degrees of standardisation. Some functions are standard, with no variants being offered. For example, the engines in an aircraft are probably not subject to negotiation.

Other functions are standard, with some variants being offered. For example, the position of the doors in an aircraft may be customer specific in terms of a few predefined variants. Still other functions could be offered with several variants, but with the additional possibility of customer-specific design. In an aircraft, this could hold for the galley. Finally, there may be functions that have to be customer-order-driven designed, such as the aircraft's interior.

A suitable information system for engineer-to-order production should be able to support the engineering activities outlined in the list above. It should be able to generate a project network for each customer order, specify deliverables at various milestones, estimate costs, etc.

In addition, the information system should draw a careful distinction between customer-order-specific information and customer-order-independent information. Customer-order-independent information should contain the latest product standards, the accepted solutions and be the basis for **generating** many (customer-order-specific) engineering documents. Customer-order-specific information may contain many elements which should never enter the standards and it should be frozen after the order has been finished. The best ideas may be retained for future reference purposes as in the make-to-order situation, but they should be investigated in depth before they can be taken into consideration for inclusion in the standards. Clearly, the information system should support this functionality.

The information system in engineer-to-order companies should support the existence of incomplete and sometimes inconsistent or ambiguous information during the design phases. Incompleteness is clear: after all, complete documentation is the result of design. Inconsistency or ambiguity arises if some engineers make assumptions about the outcome of other engineering activities, although they know that these assumptions will turn out to be false. In such situations it is far more important that an information system should be able to detect and support a temporary inconsistency than that it should force consistency. The same holds for ambiguity.

6.5 SUMMARY AND CONCLUSIONS

The aim of this chapter has been to introduce a variety of customer driven manufacturing situations. This variety has been placed in a two-dimensional grid, which constitutes a typology. For some of these types, production management issues were discussed in section 6.3. It was concluded that an upstream customer order decoupling point causes top management to focus more on innovation first and subsequently on individual customer orders. Uncertainty becomes less quantitative and more qualitative. Complexity goes upstream as well.

Section 6.4 devoted attention to information systems. Transaction processing systems may focus on products, on resources or on work flows, or any combination of these. Furthermore, when the customer order decoupling point is moved upstream, the importance of decision support decreases and is replaced by the intelligent, rule-based support of document generation and by generative solutions to production information. This will be elaborated in Chapters 16, 27 and 29.

REFERENCES

1. Hoekstra, S., and Romme, J.H.J.M.,
   Towards an integral logistics structure.
   Kluwer, Deventer (NL), 1989 *Dutch.*
2. Hegge, H.M.H.,
   Intelligent product family descriptions for business applications.
   Ph.D., Eindhoven University of Technology, 1995.
3. Orlicky, J.A., Plossl, G.W., and Wight, O.W.,
   "Structuring the Bill of Material for MRP", *Production and Inventory Management*, Vol. 13, no. 4 (fourth quarter), 1973, pp. 9-17.
4. Veen, E.A. van,
   Modelling Product Structures by Generic Bills of Material.
   Elsevier, Amsterdam, 1992.

7

Overview of the book

7.1 INTRODUCTION

Many authors contributed to this book. Each of them will have a particular *view* with regard to the design and re-design of customer driven manufacturing systems. In Chapter 2 it has been explained how these different views can be interrelated by means of performance indicators. Each author describes design choices from his viewpoint and explains how they will affect certain performance indicators that are relevant from this point of view. Different authors will, however, sometimes use the same set of closely related performance indicators. Obviously, this illustrates the need to design or re-design a manufacturing system from different viewpoints, a number of which are presented in this book.

The selection we made of the subjects to be included in this book is based on two arguments. Firstly, we wanted to cover the whole area of customer driven manufacturing. The typology in the previous chapter describes the differences between various customer driven manufacturing situations. Customer driven manufacturing covers assemble-to-order, make-to-order and engineer-to-order. A distinction can be drawn within these situations between product-oriented, process-oriented and capability-oriented situations. Given the limitations of the size of a book, it is not possible to cover each of the cells in this matrix in full detail. Each chapter will therefore focus on a limited number of cells. Our selection was made in such a way that the material covers most of the matrix.

Secondly, we needed to choose subjects that are most relevant to the field of customer driven manufacturing. Again, the number of subjects is virtually unlimited and we made a choice. This choice was based on many discussions held with the authors and many others over the last five years at conferences, workshops and, in particular, various research projects involving both academia and industry. We

76   *Overview of the book*

feel that the main issues of customer order driven manufacturing are covered by this book.

7.2 STRUCTURE OF THE BOOK

7.2.1 Design orientation

This book should serve as a guideline for engineers, managers or consultants to design and re-design manufacturing systems to achieve or improve customer orientation. It is structured in such a way that it will support this design-oriented approach.

The design-oriented approach we use is based on the control paradigm, which originates from systems science. The control paradigm distinguishes three subsystems: the controlled system, the control system and the environment (figure 7.1). In manufacturing, the controlled system is often called the primary process and the control system the production control system. The combination of the primary process and the control system is called the manufacturing system.

The information system is not explicitly mentioned in systems science since it is considered to be part of the control system. Consequently, the design of information systems is often implicitly included when discussing the design of control systems. Because of the importance of information systems in today's manufacturing, we would like to address the design of information systems more explicitly in this book. We are therefore using a slightly modified version of the

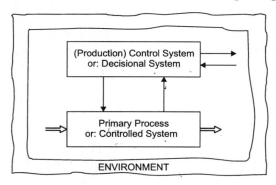

Fig 7.1   The control paradigm

control paradigm presented in figure 7.2. We distinguish between the primary process, the production control system and the information system. The book is structured accordingly.

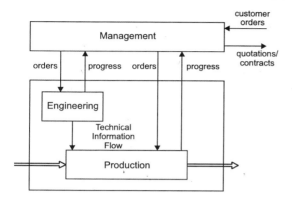

Fig 7.2 Modified control paradigm

This book is divided into 6 parts. Part A describes the background to the book: both the theoretical background of customer driven manufacturing and the economic need for customer driven manufacturing are discussed. Part B describes four cases which are used throughout the book as examples for the design and re-design of manufacturing systems. These cases cover a wide range: a company manufacturing packaging machines, which represents engineer-to-order in product-oriented manufacturing; a company manufacturing medical equipment, which represents assemble-to-order in product-oriented manufacturing; a company producing paper, which represents make-to-order in process-oriented manufacturing; and a shipyard representing engineer-to-order in resource-oriented manufacturing. Parts C, D, and E will then describe the design and re-design of the primary process, the production control system and the information system. The following subsections will discuss these parts in more detail. Finally, part F describes examples of models that can be developed for the simulation of particular design choices and performance indicators that might be relevant in a particular situation. It is argued here that simulation and modelling provide excellent possibilities for evaluating different design choices and performance indicators from different views before major investments have to be made.

7.2.2 Primary Process

In a design-oriented approach towards manufacturing systems it is important to consider first the primary process, since the characteristics of the primary process, including product characteristics, determine the need for a particular control system and information system. The primary process involves the transformation of goods and information into a product. In customer driven manufacturing this transformation will comprise engineering, work preparation, production, assembly, as well as distribution, customer service, tendering, order capture and product and process design. The means for this transformation are both machines and human resources.

In this book we consider the four major issues of the primary process in customer order driven manufacturing. These are:

- engineering (chapter 14)
- production (chapter 15)
- product modelling (chapter 16)
- human resources (chapter 17).

An important issue in customer driven engineering is the extent to which a product is designed for a specific customer or for a generic product. The costs involved with the former can be related directly to a customer. However, it will not contribute to any other customer directly. The question in the latter situation is both how to assign costs to specific customer orders and what products need to be developed. The complexity in customer driven production often lies in the flexibility needed for the production of low volume and often one-of-a-kind products. An important technique in the design of generic products is based on the concept of generic bills of material. It will allow a manufacturing company to develop a very broad range of product families without having to specify all the details of every single product.

Finally, due to the wide range of products to be produced and the flexibility required in the production process, there will be a limit to the extent a customer driven manufacturing company can automate its manufacturing. Human resources become important here. This book will address a number of issues related to human resource management, such as evolutionary human resource management, complex resource units, holistic work design, healthy informal organisations and concurrent mental modelling.

7.2.3 Production control system

The general nature of organisation and decision making in production management is discussed in Chapter 18. In production control we draw a distinction between aggregate control (including master planning), factory coordination (or goods-flow control) and production activity control or shop floor control. In Chapters 19, 20, and 21 we discuss aggregate control and factory coordination for the three main types of customer driven manufacturing: make-to-order, assemble-to-order and engineer-to-order. Shop floor control will then be discussed in Chapter 23.

Two issues that are specifically related to customer driven manufacturing are the control of engineering and the relation between the company and customers and suppliers. Customer driven manufacturing requires that engineering be controlled in much the same way as production or assembly. The nature of engineering, however, is that the 'product' produced by engineering is intangible and, therefore, that the work load is difficult to determine and to control. In Chapter 22 we will discuss how engineering can be considered as a controllable unit, contributing to the production of a customer-specific product.

Essential success factors for any customer driven manufacturing company are good customer and supplier relations. Both customers and suppliers should be seen as active partners in the manufacturing process, both organisationally and info-technically. Moreover, within the context of an all-embracing concurrent engineering effort the partnership can be extended over the entire product life cycle. The main design choices concerning customer and supplier relations will be discussed in Chapter 24.

7.2.4 Information system

The use of information technology (IT) in manufacturing is a means to achieve effective and efficient production and production control. It should therefore be based on the characteristics of the primary process and the production control system, and it will be developed accordingly. There are five areas in customer driven manufacturing that will often require specific attention when considering the design or re-design of the information system. These are:

- product modelling
- work-flow management in engineering
- quality monitoring in documentation
- software packages
- shop floor control.

For many customer driven manufacturing companies it is virtually impossible to maintain the enormous variety of products and product families without the help of information technology. The support of product modelling by automated bills of material is a well-known area for the application of IT in manufacturing. Most bill-of-material processors, however, focus on make-to-stock and assemble-to-order only. Generic bills of material are a new and powerful instrument for a mixture of customer driven manufacturing systems. The use of generic bills of material is discussed in Chapter 26.

The software market for production planning and control in manufacturing is largely dominated by MRP systems. Chapter 27 explains that this system is mainly suitable for repetitive manufacturing only. The specific requirements in different customer driven situations will be discussed in detail.

In repetitive manufacturing it is easy to define quality standards for products since the same product will be produced over and over again. In customer driven manufacturing with small batch sizes and even one-of-a-kind production a number of difficulties in documentation arise. These are discussed in Chapter 28.

As discussed in the previous section, the control of work in engineering is difficult. Chapter 29 describes how information technology can support this control by means of work-flow management.

Shop floor control too requires a slightly different approach in customer driven manufacturing. The emphasis in shop floor control in repetitive manufacturing lies on the automation of planning and control, thereby eliminating expensive and 'unreliable' human operators from the shop floor. Due to the requirement of high flexibility in customer driven manufacturing, the involvement of human operators in these situations is indispensable. A number of technical issues related to shop floor control in customer driven manufacturing are discussed in Chapter 30.

7.2.5 Models

The book concludes with six chapters on computerised models that have been developed as part of the ESPRIT basic research action 3134 "Factory of the Future." These models show how partial models can be built to support the design of a customer driven manufacturing system from a specific point of view. Chapter 32 provides an introduction to a tool that can be used for browsing through and for the design of a network of performance indicators and design choices. Chapters 33, 34, 35 and 36 will then discuss simulation models for certain design choices and performance indicators from a specific point of view. Finally, a tool for the analysis and comparison of different cases (companies) will be discussed in Chapter 37.

Part B
CASES

8

Introduction to the cases

The second part of this book presents four cases. These cases have been selected in such a way, that they cover to some extent the typology presented in Chapter 6. The cases serve several purposes.

First of all, the cases should help the reader to understand the large differences which may occur in customer driven manufacturing. As we proceed through this Part B, this point will certainly will be appreciated.

The four cases are described in terms of products, resources, workflow, internal organization and decision structures. This description is in line with the concepts introduced in Chapter 5 (the so-called theoretical framework). Furthermore, these cases are described in a way which underscores the strategic design choices made by the companies involved. In the remaining Parts of this book, we will use these cases to discuss consequences of these design choices, alternatives, and presumed relationships with performance indicators.

The first case to be discussed (in Chapter 9) is concerned with a company that produces packaging machines. This company has developed various packaging principles and technologies, but each machine requires a considerable amount of additional engineering. Therefore, the company is a product-oriented engineer-to-order company. The case description highlights especially the nature of customer driven engineering. In fact, the case is used as basis for a discussion on:

- the organization of customer driven engineering in Chapter 14, Part C
- the management of tendering and engineering in Chapter 22, Part D
- work-flow modelling in engineer-to-order information systems in Chapter 29, Part E
- a model relating design choices to performance indicators with respect to customer-order acceptance in Chapter 36, Part F.

86   *Introduction to the cases*

Some of these topics are discussed in more detail in Muntslag (1993). The second case, which is discussed in Chapter 10, describes a company which produces medical equipment. The company is on its way to adopt a clear product-oriented assemble-to-order strategy. Although this company has its product orientation in common with the previous case, there are large differences. For example, customer-driven engineering is absent in this case. As another example, the role of product modelling techniques is much more important here. Furthermore, the appropriate logistics concept for this company is influenced by the company's customer orientation as reflected in field installation policies, after sales service requirements, etcetera. Therefore, this case is used as basis for a discussion on:

- product modelling in customer-driven manufacturing in Chapter 16, Part C
- production planning & control in assemble-to-order manufacturing in Chapter 20, Part D
- product modelling and information technology in Chapter 26, Part E
- document management in Chapter 28, Part E
- a model relating design choices to performance indicators with respect to Production Planning in Chapter 36, Part F.

Some of these topics are discussed in more detail in Hegge (1995) and Erens (1996).

The third case, which is to be discussed in Chapter 11, deals with fine paper manufacturing on customer order. The case is workflow oriented, although elements of resource orientation are also present. This case is a make-to-order case, and the production plannning & control concept is most crucial here. In a paper factory, the physical layout and organization of the shop floor play an important role. Therefore, this case constitutes a basis for the discussion on:

- the organization of production in Chapter 15, part C
- production planning & control in workflow-oriented manufacturing in Chapter 19, part D
- shop floor management in Chapter 23, Part D
- shop floor control (information) systems in Chapter 30, Part E
- a model relating design choices to performance indicators with respect to shop floor management in Chapter 35, Part F.

The fourth and last case (Chapter 12) is concerned with shipbuilding. This is a resource oriented case, which shows the importance of not only physical resources such as cranes and halls, but also of human resources and software. This case provides another basis for the discussion of production in customer driven manufacturing (Chapter 15) which is complementary to the fine paper case. The shipbuilding case provides also the basis for the discussion on resources, which is the subject of Chapter 17, Part C. Moreover, the case serves as an illustration for Chapter 21, discussing production planning & control in customer-driven manufacturing.

Also, the shipbuilding case provides a good example of an extended enterprise. Therefore, the case also provides inspiration for the discussion of this subject in Chapter 24 of this book. Finallly, Part F contains two models for Human Resources Management design choices which are based on the case of Chapter 12. These models are described in Chapters 33 and 34.

REFERENCES

1. Erens, F.J., *The synthesis of variety*, Ph.D. Thesis, Eindhoven University of Technology, 1996.
2. Muntslag, D.R., *Managing Customer-order driven engineering*, Ph.D. Thesis, Eindhoven University of Technology, 1993.
3. Hegge, H.M.H., *Intelligent product family descriptions for business applications*, Ph.D. Thesis, Eindhoven University of Technology, 1995.

9

Engineer-to-order in product-oriented manufacturing

9.1 INTRODUCTION AND COMPANY DESCRIPTION

In this chapter we describe a company which manufactures custom-built packaging equipment for industrial customers. This company is a typical product oriented engineer-to-order company (see Chapter 6). This company specializes in the manufacture of packaging equipment for three application areas, namely:

- the (primary) packaging of liquids (not containing CO_2);
- the (secondary) packaging or boxing of a large range of prepackaged products;
- the (tertiary) packaging or loading of boxes onto pallets.

Before the detailed characteristics of the organization and the transformation process are analyzed, a summary of the selected case situation is presented in this section in terms of several general characteristics, the range of products and the sales market. For further details, see Muntslag (1993).

9.1.1 General characteristics

The company supplies single machines as well as complete production lines with multiple packaging machines to its customers. The annual turnover is between 55 and 60 million guilders, of which more than 90% is derived from the sale of single units. The added value contributed by the company through its products averages approximately 45%. The delivery lead time for custom-built units varies from three to nine months, depending upon the extent of the product specifications provided by the customer. Approximately 240 persons are employed by the company, of which about half are directly involved in the production process.

9.1.2 The range of products

The range of products is divided into three product families or product groups, namely: primary packaging equipment (bottling machines), secondary packaging equipment (packaging machines) and tertiary packaging equipment (palletizing machines). Each of these machines consists of a large number of components (i.e., several thousand). More of the components in packaging machines and palletizing machines are typically customized, leading to a greater percentage of customer order driven engineering. The installation of bottling machines is relatively labor-intensive. The flexibility with respect to sharing resources between product groups is limited.

9.1.3 The sales market

The company supplies equipment world-wide to manufacturers of fast moving consumer goods. These manufacturers are found primarily in the more developed countries and highly populated metropolitan areas. Equipment is supplied to companies in a variety of industrial sectors. The company has an enviable reputation for manufacturing high-quality equipment. The average delivery lead time in the market varies from three to nine months, depending upon the size of the order and the degree of customization.

9.2 THE INTERNAL ORGANIZATION (RESOURCES VIEW)

The internal organization is divided into three functional divisions, namely: Sales, Engineering and Production. In addition, there are three staff departments: Finance, Organizational Development & Data Processing and Human Resources (see figure 9.1). Each of the three functional divisions is described in more detail below.

The interal organization (resources view) 91

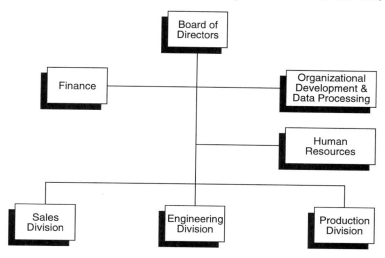

Fig. 9.1 High-level organization structure of the company

The sales division

The organizational structure within the Sales Division is relatively complicated. The total market area is segmented geographically into four Areas. Each Area is represented internally by one or more Area Coordinators (ACs). The sales channels consist of sales offices in the four primary sales countries and local agents in the other countries. There is no segmentation by product group within the Sales Division; each Area and each sales channel is, in principle, responsible for selling products in all three of the product groups. A total of approximately 50 persons are employed in the Sales Division.

The engineering division

Unlike the Sales Division, the Engineering Division is organized primarily along the lines of the product groups. A distinction is made between Bottling Technology and Packaging Technology within the Engineering Division. Both the Bottling Technology and the Packaging Technology sections are split up functionally into a Product Engineering Department where the product concepts are initially developed, and a Detail Design Department in which the subsequent realization of the product design is worked out. The Engineering Division also has a separate Control Department for developing the machine control functions. Approximately 55 persons are employed within the Engineering Division (refer to figure 9.2).

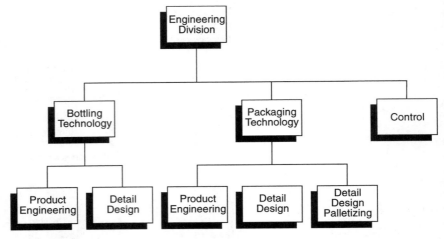

Fig. 9.2 Organization structure of the Engineering Division

The activities associated with the customer order driven engineering fall primarily under the responsibility of the Engineering Division. Much of the analysis included in the remainder of this chapter is therefore related to the activities within this division.

The production division
The Production Division consists of four departments which are directly involved with production activities and four other departments which are indirectly involved. The component manufacturing activity is split up into three production departments, namely: the Milling Shop, the Stamping Shop and a Universal Group. This last department is primarily responsible for manufacturing spare parts in connection with after sales service. In addition to component manufacturing, there is also an Assembly Section within the Production Division. The four departments which are indirectly involved with the production activities are: Purchasing & Production Control, Process Planning & Technical Support, Quality Assurance and Technical Documentation & Administration. Approximately 120 persons are employed within the Production Division.

9.3 THE TRANSFORMATION PROCESS (WORKFLOW VIEW)

The company described in this chapter can be characterized as a typical engineer-to-order plant. This means that the transformation process can be subdivided into a non-physical production process, the customer driven engineering and a physical production process consisting of component manufacturing and product assembly. The way in which both of these sub-processes are carried out in the present situation, is described below.

9.3.1 Customer order driven engineering

At the highest level, the customer order driven engineering sub-process can be split up into three phases, namely: the sales lead phase, the quotation phase and the order phase.

Sales lead phase
The sales lead phase is initiated when an initial contact between a potential customer and one of the sales channels (agent, sales office or AC) takes place. It is assumed that the customer has a problem and is interested in finding out if the company can provide a solution. The salesman forms an opinion about the customer's requirements and determines to what extent a solution can be provided. If the salesman believes that there is a good chance of winning the order, then the customer's requirements are documented using a standard sales order checklist. This request for a detailed quotation is then passed on to the responsible AC who gives this to the Engineering Division. This initiates the formal quotation phase.

Quotation phase
The request for quotation is received by the Engineering Division and then evaluated and processed by a product engineer within the Product Engineering Department. The product specifications are only worked out in general terms at this stage in the form of a general technical description. This document is a summary of the customer requirements translated into product functions and provides the technical content for preparing a detailed quotation. After the price is established and a delivery date is determined, the quotation is then sent to the customer. Negotiations with the customer can also result in changes to the price or the delivery lead time. One or more of the quotations will ultimately lead to either a

customer order or a breakdown in the negotiations. If the customer decides to place an order, then the order phase is initiated.

Order phase
After the existence of a new customer order is made known within the organization, the Engineering Division works out the order specifications in more detail. The product engineer takes the general product specifications and fills in the details on technical order sheets. These technical order sheets are then passed on to the detail designers (in the Detail Design Department) who are specialized in the realization of the final product. These detail designers add the technical drawings, bills-of-materials and further details to the product specifications. The Detail Design Department then defines various *installation units* for simplifying the installation of each custom-built machine. Each installation unit can be seen as a group of components and sub-assemblies of the machine which are to be assembled simultaneously. A separate assembly order is drawn up for each installation unit. The specifications for each installation unit are then worked out separately, ensuring that the sequence of detailing is coordinated with the assembly sequence of the installation unit. As soon as the detailing activities for an installation unit have been completed, these are released to the Production Division.

The customer driven engineering process is finished when all of the product specifications have been completed. Manufacturing instructions first need to be prepared, to the extent that these instructions are not already available, before the physical manufacturing of the customized components can be initiated. These last activities are carried out within the Process Planning Department (of the Production Division).

9.3.2 The physical production process

The physical flow of goods consists of the physical production and purchasing activities. A significant portion of the components are purchased; even whole machine modules, often based upon customer specifications, are sometimes purchased as a part of a customer order. Purchase orders for components and sub-assemblies, as well as the arrangements for outsourcing of production capacity, are coordinated via the Purchasing & Production Control Department. The components and simple sub-assemblies which are neither purchased nor outsourced are produced internally within the three functionally-oriented manufacturing departments. The final customized product is assembled within the Assembly Section. After testing, the machine is partially disassembled and then shipped to the customer. The On-site Installation Department (of the Sales Division) is responsible for the final installation and start-up at the customer's plant.

9.4 PRESENT MANAGEMENT OF THE TRANSFORMATION PROCESS

Attention is paid to the production control (timeliness factor), in the first place. In the second place, controlling the quality and cost factors is described in further detail.

9.4.1 The present method of production control

Three planning levels can be identified for production control within the company, namely: operational planning (independent of specific customer orders), first-cut planning and detailed planning (customer order dependent).

Operational planning
An operational plan is established once each year. This plan includes a forecast of the product quantities expected to be sold during the year for each product group. The medium-term resource requirements can then be calculated based upon this forecast. In this way, the operational plan represents an annual, coordinated effort (at a high level) to match the sales plans with the production activities. The forecasts included in the operational plan are also used as the basis for an annual determination of the standard hourly rates used in calculating the product cost prices (see further).

First-cut planning
Standard delivery lead times are used as the basis for customer quotations. The standard delivery lead time is determined based upon the number of hours estimated for detailed design and the finished product(s) to be supplied. In view of the fact that the throughput time for the quotation phase can be fairly unpredictable, the future workload of the various resource capacity groups is taken into account only minimally. A first-cut plan in the form of a project network with aggregated activities is created for each customer order received. An

estimate is made of the required resource capacity per group of aggregated activities based upon the general technical description. The earliest and latest start dates and completion dates are calculated per activity. Standard project network structures have been defined for each product group. Internal due dates are calculated based upon the workloads within the Detail Design departments. Each customer order is added to the internal production schedule of the relevant Detail Design Department. The scheduled start date for the physical production is determined solely based upon the internal throughput time of the detailing activities in view of the limited flexibility and the limited diversity of skills found in connection with the detailing resource capacities. The due date is calculated by adding the standard throughput times for manufacturing and assembling the components to the scheduled completion date for the detailing activities. Capacity utilization profiles are then developed for each of the Detail Design departments, the Manufacturing departments and the Assembly departments based upon the project networks which have been established. Whenever a profile shows a planned over-utilization of the available resource capacity in one or more of the capacity groups, an effort is made to utilize the slack in the various project network structures in order to resolve the capacity shortage. If this does not provide a solution, then other measures such as overwork, outsourcing and reassignment of multi-skilled employees are taken to expand the available capacity. As a last resort, the scheduled due date for the order is postponed and the customer is notified of the anticipated delay in delivery.

Detailed planning
No detailed plans and priorities are used routinely within the Product Engineering and Detail Design departments. The ACs approach specific individuals within the Product Engineering Department to arrange for the preparation of quotations. Priorities are, thus, set on an individual basis. The activities carried out by the Detail Design departments follow directly from the first-cut planning. A relatively simple human resource plan is also maintained to keep track of which persons are involved in which projects. The priority is determined based upon the high-level plan made for the installation units within the assembly phase. After the detailed design and the process planning has been completed, detailed planning is performed for the production of components and the assembly; the coordination and requisitioning of materials is also initiated at this point. An integrated, automated production control system is used for this purpose. A significant number of standard components are also made for stock, driven

independently of the customer orders by the stock levels. A weekly summary of the order progress is prepared for each customer order. This summary provides a report of which departments have completed their tasks for each of the installation units.

9.4.2 The present method of financial control

The present method of financial control over customer orders is concerned primarily with monitoring the actual project costs. This is done by preparing a project budget, keeping track of the actual project costs and then comparing the actual versus budgeted costs. The full cost price of the future product is calculated based upon the general technical description which was prepared during the quotation phase. This is calculated by the product engineer in most instances, but sometimes by a technical budgeting specialist. The full cost price is determined by estimating the material costs for each "product function" specified in the general technical description and adding to this the costs of using the specified resource capacities. The full cost price for the complete machine is obtained by finding the sum total of the full cost prices for all of the relevant product functions. The quotation price is calculated by the Sales Division and consists of the full cost price plus the desired profit margin.

If a quotation is accepted and becomes an actual order, then the project budget calculation which was made during the quotation phase is revised and finalized as a financial target for the organization. The actual hours per type of resource capacity and the actual costs of materials are subsequently accounted for per project. A comparison of the budgeted costs and actual costs can be made based upon this information for each customer order. This analysis is carried out after the product has been delivered to the customer and the customer order has been completely processed and closed.

9.5 FURTHER ANALYSIS OF THE PROBLEM ISSUES

Based upon an extensive analysis of the selected case situation, a number of important problem areas have been identified in connection with the process of customer order driven engineering. These problem areas are described in more detail in this section. The problem issues are first summarized, below. This is followed by a detailed discussion of these issues.

9.5.1 Summary of the current problem issues

The current internal and external performance in the selected case is far from optimal. This is apparent with respect to:

- uncontrolled product quality, evidenced by the large number of engineering changes which take place after the engineering drawings have been released. In addition, an extremely large number of quality problems are discovered during the internal product assembly phase and when the product is installed at the customer site;
- uncontrolled internal throughput times for the customer order driven engineering as well as for the physical production processes. This results in a situation in which over 60% of the customer orders are delivered too late;
- uncontrolled customer order costs, evidenced by the large discrepancies between the budgeted costs and actual costs as well as profit margins which are too low.

A significant number of the problems described here appear to be associated with the non-physical customer driven engineering phase. In view of this, an in-depth analysis of the connection between the problems found in the customer order driven engineering and the problems concerning quality assurance, production control and cost control for the organization as a whole is used as the starting point for the redesign in subsequent chapters. The results of this analysis are presented in the sub-sections below. In connection with this, a distinction has been made between the problem issues related to the engineering process, the control of the engineering process and the provisions for management information.

9.5.2 Analysis of the engineering process

The most important problems concerning the current customer driven engineering process are analyzed in this sub-section. This provides an input for further chapters in this book.

The structure of the customer driven engineering process

The most important problem concerning the current engineering process in the selected case is the absence of a proper structure for this engineering process. Several different aspects are involved. In the first place, a formal engineering approach is missing. The use of a formal approach implies, for example, that a product is developed following a certain method which takes the customer-specific requirements into account and translates the specified functions into a tangible form via a series of formal engineering steps (Pahl and Beitz 1988, Roozenburg and Eekels 1991). This type of formal, structured approach is not used within the current engineering process. The current engineering approach can be characterized as "engineering based upon previous products". Customized products are engineered based upon designs which have been used in the past and approximate (to some extent) the same functionality. Each product engineer and detail designer is free to contribute his own interpretation of how the product is to be developed. This problem is complicated by the fact that the documentation related to previous orders is poorly organized and poorly indexed. A formal description of the current engineering method also does not exist. This means that quality assurance is virtually impossible and that the training of new employees is hindered due to a lack of documented procedures.

In the second place, each product engineer and detail designer has his own way of documenting the product specifications. For example, there is no formal approach for developing engineering drawings and there are no naming conventions for identifying components. As a result, a multitude of problems may arise with respect to different interpretations of the product specifications.

In the third place, the coordination and communication between the various disciplines during the customer driven engineering process is poor. This is true with respect to the transfer from Sales to the Product Engineering Department as well as the transfer from Product Engineering to the Detail Design Department. The coordination activities and the communications about customer orders involve substantive information only when (acute) problems arise. This lack of coordination and communication can be seen as a direct result of the absence of a formal engineering approach and formal engineering documentation. This means that each discipline tends to develop its own standards and documents for use within that particular department rather than for the effective communication of information to other disciplines. This inhibits the effective coordination of activities and communications between departments and leads to the creation of "functional walls". Order information is then merely "tossed over the next wall" following the completion of each step in the process.

The technical engineering split in terms of installation units

Within the Detail Design Department, a custom-built machine is split into so-called *installation units*. Each installation unit represents a group of components and sub-assemblies belonging to a certain machine which are assembled together at the same time. A separate assembly order is also issued for each installation unit. Each installation unit is engineered individually, in the same sequence as the assembly sequence for the installation unit. As soon as the engineering for an installation unit has been completed, this documentation is released to the Production Division. A custom-built machine is split into installation units based upon the standard grouping of components which is pre-defined for the basic machine of this type. As such, the installation units can be seen as the highest level of the bill-of-materials. Each installation unit is modified as necessary to comply with the customized specifications. The current grouping in installation units has developed gradually over a long period of time, however, so that this now does not necessarily represent a partitioning of the product into modules which are logical from an engineering point of view. Some of the installation units have been defined as such because they are purchased from external suppliers or because they are assembled in a single operation. Nevertheless, there are a large number of interrelationships between the installation units. As a result, most of the installation units are not very modular and cannot be engineered independently, even though this is attempted in practice. The installation units associated with a given customer order are engineered and released more-or-less individually, often by different persons. In many instances this results in a situation in which installation units which have been released to the Production Division need to be recalled for modifications. This means that there may be a significant flow of (rush) orders for engineering changes for components which are already in production. This has a serious negative effect on the quality, financial result and timeliness of the customer orders.

Customer driven and customer independent engineering

Within the Engineering Division, no sharp distinction has been made between customer driven and customer independent engineering. Customer independent or innovative engineering is generally carried out based upon specific customer orders in the current situation. Whenever there is a requirement for a new function, a new product module or even a new product family, the current policy is to wait until a suitable customer order has been received which can then be used as the basis for the new development. The advantage of this approach is that the customer pays, at least partially, for developing the innovation in this way. This approach also has a number of major drawbacks, however:

- the technical risk is relatively high due to the innovative character of the development. This means that significant financial and time risks could also be present. This type of customer order can, in addition, always be characterized as being difficult to control;
- the basic nature of customer independent engineering is different from customer driven engineering. Since it is intended that the first type of engineering should lead to the establishment of future product standards, it is important that innovations in this respect are properly developed and tested with respect to the technological and financial (cost price) feasibility. Characteristic of this type of engineering is the attitude that "it can always be improved". This attitude is not appropriate in connection with customer driven engineering where customized developments are generally one of a kind and the motto should be "good is good enough". If customer independent engineering is linked to actual customer orders, then there is a real danger that insufficient attention will be spent on the objectives which are independent of the customer order.

9.5.3 Analysis of the controlling system

The most important problems associated with the current controlling system for the customer driven engineering process as analyzed in this section provide input for further chapters of this book.

The formal controlling system
A formal controlling system for customer driven engineering is absent in the current situation. This is apparent with respect to a number of aspects. In the first place, formally defined control decisions and the responsible decision-makers have neither been defined nor identified. Control decisions in this sense should be found with respect to, for

example: issuing quotations, setting prices, accepting orders and releasing work orders. Since these decisions have not been formalized, the related responsibilities and delegated authorities within the organization remain vague and ambiguous. In practice, this means that many of the control responsibilities are carried out, and priority decisions made, at a level which is (too) low in the organization by persons who would normally not be authorized to make such decisions. Given this ambiguity concerning responsibilities and authorities, no one feels responsible and it is difficult to hold anyone responsible for a specific order result.

In the second place, a formal review of the requests for quotation is not carried out. Since it is not possible to respond to all of the requests for quotation which are received during a given period of time, a method is needed for prioritizing and accepting such requests. This should be based upon criteria which are important to the company such as the product/market strategy. This prioritizing is not done in the current situation. Requests for quotation are accepted or rejected by various individuals within the sales organization and priorities are established based upon individual discussions with the product engineers. An explicit risk analysis concerning the requests for quotation is also absent. No analysis is made in advance of the existing uncertainties which imply certain risks. As a result, unpleasant surprises often occur in a later phase. This type of risk analysis could provide important input for the acceptance and establishment of priorities for processing the requests for quotation. Since no risk analysis is currently performed, no distinction is made between the requests for quotation which would fall into different risk categories.

Calculating a full cost price

In the current situation, the quotation prices are determined in part by calculating a full cost price based upon average hourly rates with surcharges for allocating the indirect costs. A similar method based upon multiple surcharges (Horngren and Foster 1991) is often used in practice, even by larger companies which produce a wide range of different products (Howell et al. 1987). In view of the simplicity of this approach and the method used for allocating costs, however, there is little causal relationship between the indirect costs and the product at hand. In addition, the actual control effort and the actual use of indirect resources may vary radically between product groups as well as between orders in view of the diversity of products (particularly between product groups) in the selected case situation. Use of the current method of calculating cost prices and the "arbitrary" allocation of indirect costs means that certain product groups may be burdened with excessive costs while a fair share of these costs is not allocated to other product groups (see also Cooper and Kaplan 1987). The allocation of *direct* costs is, in fact, also imprecise due to:

- the lack of a formal engineering approach which implies that estimates of the material costs generally will be inaccurate;
- the lack of a risk analysis which implies that a reliable estimate of the required hours per type of direct resource capacity cannot be made.

9.5.4 Analysis of the management information resources

It is not surprising to discover that there is little evidence of an integrated management information system in the selected case situation in view of the absence of a formal controlling system for customer order driven engineering. Several local information systems have been established, however, these systems are not integrated. The most important systems are:

- a quotation registration system in which each receipt of a request for quotation is logged and the issued quotations are registered;
- a (financial) project administration system in which project costs are entered for the purpose of keeping track of the project investments. This information is of limited use for the (cost) control of orders in view of the financial nature of the data, the delay in entering and processing the data in the system and the budgeting structure which is used (see previous discussion);
- a network planning system in which the project network structures for potential customer orders can be defined in order to generate resource capacity requirement profiles. This information is then used to determine final delivery due dates. This network planning system is completely separate from the management information system for the physical production process, however. The network planning system is also used to track the progress of aggregated activities (at a high level) based upon summary data provided by the various departments.

In the current situation, there is little systems support for the registration of data to monitor the status and progress of customer order driven engineering activities. Status information is collected on an ad hoc basis (primarily verbally) and generally consumes a significant amount of time.

CONCLUSION

The problems discussed in this chapter for the product oriented engineer-to-order case will be discussed in several subsequent chapters of this book. Chapter 14 describes how engineering can be organized as customer driven manufacturing. Chapter 21 presents a production control system. Chapter 22 describes a management approach for tendering and engineering. All these chapters refer back to the case just described. For further reading, see Muntslag (1992).

REFERENCES

1. Cooper, R. and R.S. Kaplan, How cost accounting systematically distorts product, in *Accounting & Management: Field Study Perspectives*, W.J. Bruns and R.S. Kaplan (eds.), 1987
2. Howell, R.A., J.D. Brown, S.R. Soucy and A.H. Seed III, *Management accounting in the New Manufacturing Environment*, Montvale, 1987.
3. Horngren, C.T. and G. Foster, *Cost accounting: a Managerial Emphasis*, Prentice-Hall, 1991.
4. Muntslag, D.R., *Managing Customer Driven Engineering*, Doctoral Thesis, Eindhoven University of Technology, Eindhoven, 1993
5. Pahl, G. and W. Beitz, *Engineering design: a systematic approach*, Design Council, 1988
6. Roozenburg, N.F.M. and J. Eekels, *Produktontwerpen, structuur en methoden*, Lemma (in Dutch), 1991

10

Assemble-to-order in product-oriented manufacturing

10.1 INTRODUCTION

In this chapter we describe the case of a manufacturing company, Medicom, which specialises in the manufacture of medical equipment for professional use by hospitals and doctors with private practices. Most of Medicom's medical systems can be supplied in more than a million different configurations (product variants). This, together with a demand lead time of half a year, provides the basis for Medicom to configure and assemble products specifically made to customer order.

We will explain, however, how Medicom is operating as if it is in an engineer-to-order situation in the way it develops and adapts products for individual customer orders. In response to the problems caused by this, the Management of Medicom has decided to introduce for some products a totally opposite approach by configuring and manufacturing a very limited set of product variants.

With this approach, Medicom has now adopted two different specification schemes and related production control approaches, each with very serious drawbacks. In this chapter we will describe this situation in more detail. Subsequently, in Chapter 16, a possible approach is presented for resolving Medicom's configuration problems. Chapter 20 discusses production control. Finally, in Chapter 26, we will describe how information technology can be used to implement this solution. For further details, see Erens (1996).

10.2 COMPANY DESCRIPTION

Medicom is one of the world's largest producers of medical X-ray equipment. Although it is a subsidiary of a major (Japanese) corporation, it operates independently from its parent company in the various world markets. Medicom is particularly interesting as a case study in view of the fact that the company has already evolved through several phases of development with respect to customer driven manufacturing and managing of product variety. Section 10.2.1 provides a brief summary of the company structure. Section 10.2.2 focuses on some of the recent changes.

10.2.1 General characteristics

Figure 10.1 shows the basic organisational structure of Medicom. The most important organisational aspect of Medicom is that a clear distinction is made between:

- System Operations (responsible for manufacturing finished products)
- Component Operations (responsible for manufacturing subsystems and components)
- Commercial Operations (responsible for sales and service)

The first functional area is relatively new within Medicom. Previously, products manufactured by Component Operations were sold as a set of subsystems and components. This subject is discussed more extensively in Section 10.5.

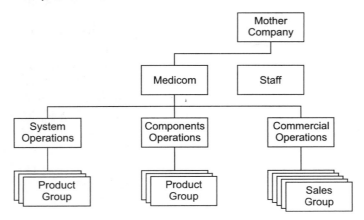

Fig. 10.1 Basic organisational structure of Medicom

The following list contains examples of some of the Product Groups found in the first two functional areas.

Component Operations
- image intensifying
- digital image manipulation
- power generation
- patient support

System Operations
- bucky systems
- surgery systems
- radio fluoroscopy systems

Figure 10.2 illustrates that the Product Groups are responsible for product development, marketing and manufacturing of components/subsystems or finished products. The Sales Groups are responsible for all of the direct relationships with customers, i.e., sales and service.

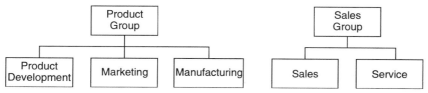

Fig. 10.2 Product Groups and Sales Groups

Worldwide, there are over 10,000 Medicom employees in 12 Product Groups (6 countries) and 20 Sales Groups (20 countries). These people are involved for a large part in knowledge-intensive business processes such as development and to a lesser extent in the physical production activities. With respect to the latter, Medicom focuses on the assembly of systems and subsystems. Low-level components (which are often complex products, nevertheless) are produced by a large variety of suppliers located near the factories, creating work for an additional 3000 people.

Together, the employees of Medicom are responsible for an annual turnover of 2 billion ECU. On the average, profits have been high, due to an increasing government demand for health care equipment.

10.2.2 Recent changes: components versus systems

Medicom manufactures a large variety of subsystems and systems. Most of the subsystems and components manufactured by Component Operations are found in more than one system. X-ray tubes, for example, are used in a number of different product configurations throughout the total range of systems. Other subsystems are developed specifically for a single family of finished products; in this case the development normally takes place under the responsibility of one of the Product Groups within System Operations.

Medicom started out originally as a group of manufacturing plants producing components and selling them via distributors in only a few countries. At that time the sales and service engineers had a major responsibility for designing systems which would work effectively in practical situations. Intensive discussions with the manufacturers of the components were needed to develop the necessary interfaces. Some of the development work was often carried out locally by a Sales Group.

Early in the Eighties, this turned into an undesirable situation for a number of reasons:
- the systems became increasingly complex and used more embedded software which increased the need for good interface management;
- the increasingly complex systems required more development effort and a balanced life-cycle management to increase the reusability of existing components;
- the customers demanded an integrated user interface instead of a heterogeneous set of components in which various customer functions are duplicated;
- the customers required a quick installation of a system in a hospital, assuming a first-time-right implementation without the need for developing customised interfaces;
- the customers wanted a more application oriented discussion with the Sales Groups.

The acknowledgement of these problems has resulted in the creation of Product Groups which are responsible for the marketing, development and manufacturing of the finished products. Sales Groups now order products from both System Operations and Component Operations (see Figure 10.3). This is discussed further in the remainder of this chapter.

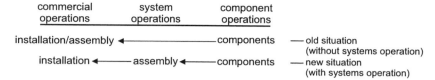

Fig. 10.3 Component Operations versus System Operations

10.2.3 Potential for assemble-to-order manufacturing

It would appear that Medicom is positioned well for assembling products to customer order. The acceptable lead time for hospitals and doctors with private practices is often more than five months since the installation of a medical system typically requires an extensive refurbishment of the examination room. On the average, only four months are actually required to assemble, ship and install a medical system. When the system can be assembled by System Operations, it is clear that most of the medical system can still be manufactured to customer order, but with the additional opportunity for reducing stocks and uncertainty.

Nevertheless, the aforementioned reasons have led to a situation in which considerable more time is needed to assemble a new system from components at the customer site.

In Section 10.5, we will demonstrate the important relationships that exist between the choice of a production control approach and the way in which products are specified and identified. In an ideal situation, the method for specifying products is derived from the production control approach. At Medicom, however, the chosen method for specifying products (by anonymous code numbers) has hindered the introduction of assemble-to-order manufacturing (ccf. Chapter 6, Section 6.4).

10.3 THE RANGE OF PRODUCTS

In this section, several observations are made concerning the range of products. To start with, a short introduction is provided to describe the general architecture of a medical system. Based upon this, it is explained how the variety of components can lead to a very large variety of finished products.

10.3.1 Product architectures

In general, X-ray systems are built from the following components/subsystems:

- a stand (to support other components);
- a generator (to generate power);
- an X-ray tube (to produce the radiation);
- an image intensifier (to intensify the radiation after the patient has been scanned);
- an image manipulation unit (to improve the image quality);
- a user console (to safely and easily initiate all of the system functions);
- a table (to support the patient).

10.3.2 Product variety

A medical system could be sold in more than a million different configurations, even though only a few hundred product variants are actually sold each year. It is impossible to predict precisely which finished product variants customers will want. This means that Medicom is forced to develop product families from which a customer can configure his/her own specific variant.

The different possible configurations of a medical system at Medicom are based upon a wide variety of options at the subsystem level. Several subsystems have been defined in terms of a limited number of basic configurations intended to suit a majority of the requirements of the individual customers. Furthermore, a single medical system architecture has been developed to accommodate all of the different subsystem configurations. Each combination of configured subsystems, thus, results in a different finished product variant. This is illustrated in Figure 10.4 where:

- the finished product variants can be found at the top of the hourglass figure;
- the various subsystem configurations are in the neck of the hourglass; and
- the components are at the bottom of the hourglass.

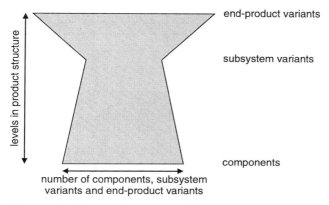

Fig. 10.4 The product variant "hourglass"

10.4 THE TRANSFORMATION PROCESSES (WORKFLOW VIEW)

Two different transformation processes can be identified: product development and manufacturing. Component development is market driven and not customer driven. Nevertheless, component interfaces are developed locally by service engineers for individual customers. The manufacturing of components is driven by planning. The following sections (10.4.1. and 10.4.2) describe the two transformation processes in more detail.

10.4.1 Product development

The development of medical systems is carried out along two lines, namely:

- development of subsystems/components;
- development of finished products (systems).

Heavy demands are placed on the development of standard interfaces between subsystems/components and on life-cycle management. This is due to the fact that subsystems and components need to be developed so that they can be used in several systems. In addition, the systems must have such an architecture which can accommodate a variety of subsystem configurations. The development phase, therefore, necessarily includes more than the development of just a single product family and involves a close interaction between Component Operations and System Operations. The customer requirements as well as the available technology for the coming years are taken into account to derive a functional model of a product family. The main processes and information flows are described in this functional model, but they are not yet mapped to physical components at this stage.

The development process eventually leads to the creation of an architectural model of the product in which functions, processes and information flows are mapped to technological solution principles. The detailed design phase then includes the preparation of drawings, compiling bills-of-material and building the initial prototypes. Also included in the detailed design phase are changes to the physical components as may be required for assembly purposes (even though design-for-assembly is not a key priority for Medicom).

The total development process often takes three years or more to complete. In order to remain competitive, however, the development lead time will need to be reduced in the future.

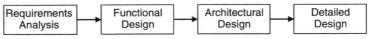

Fig. 10.5 Development phases

10.4.2 Manufacturing

Although product development is considered to be critically important to Medicom, the primary business process is still the manufacture of the actual products based upon customer orders. Assuming that System

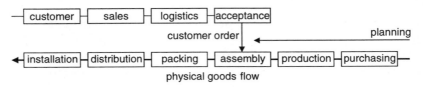

Fig. 10.6 Manufacturing process

Operations is responsible for assembling the finished products, this primary business process is illustrated in Figure 10.6.

As shown in Figure 10.6, numerous elements are involved in the manufacturing process. The customer order, for example, involves the customer, sales, sales logistics and order acceptance. The customer order driven flow of goods involves final assembly, packing, distribution and installation. Finally, production and purchasing are initiated based upon production and purchasing plans. Each function requires a specific subset of information from the customer order, varying from purely functional requirements to specific manufacturing bills-of-material. Each subset of customer order information must be appropriately translated and provided to the respective function according to the "view" it requires, without loss of information.

Although much of the lead time is used for processing the customer order (often more than a month), it is still possible to also prepare the final assembly schedule (FAS) based upon the customer order information. The components are manufactured based upon a production plan, however.

In view of the fact that the total lead time of a customer order — from signing the contract through to the installation of the product — is 4 months, there is an opportunity for further optimisation of the manufacturing process since this time is considerably less than the required lead time of 5 months. This slack time is not utilised for moving the customer order decoupling point "upstream" for a number of reasons which have to do with the methods used for specifying products. These reasons are explained in the next section.

10.5 CONTEMPORARY METHODS OF CONTROLLING THE PRIMARY PROCESS

As briefly described in the previous section, processing and translating the customer order information typically consumes a considerable portion of the total lead time. Until now, Medicom has tried two different methods of specifying and conveying the information about product variants. Both of these methods have serious drawbacks. Furthermore, both specification methods are tied in with the existing production control approaches. Experience has shown that it is not good practice to let more than one production control approach dictate how products are to be specified and identified. In the current situation, two ways of identifying product variants (corresponding with the two

114 *Assemble-to-Order in Product-Oriented Manufacturing*

production control approaches) are used simultaneously: a specification based upon components and a specification based upon the preferred systems.

10.5.1 Specifying components

Although some of the Product Groups at Medicom are responsible for manufacturing systems, most of the Product Groups are specialised in manufacturing specific components. Medicom can, therefore, best be viewed as a conglomeration of factories which manufacture components. Sales Groups generally order components at the different factories and wait until all of the components arrive at the sales organisation before the assembly process is initiated. Assembly of the product at the customer site normally takes a few months, including the time needed to develop new interfaces when the product configuration incorporates new combinations of components.

The factories are fairly autonomous; there are only a limited number of activities which require coordination across multiple factories. One of these activities is the publishing of a catalogue by the headquarters of Medicom every 6 months. This catalogue can be seen as a list of components and an indication of which components can be combined to create a working system. A sales order consists of a selection of components from this catalogue which are subsequently ordered from several different factories as may be necessary (Figure 10.7).

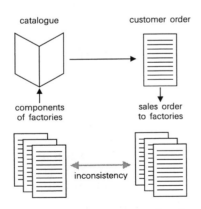

Fig. 10.7 Catalogues and orders

As indicated in Figure 10.7, the list of actual components maintained by the Product Groups may not be consistent with the components as specified by the Sales Groups in the sales orders. This inconsistency is due to:

- numerous engineering changes to components which may not be reflected in the component specification used for ordering;
- the tedious process of defining groups of components in the sales catalogue to facilitate the selling process.

As a result, Sales Groups and Product Groups have different opinions regarding when the component specification should be updated. Sales Groups take the functionality of a component as a leading criterion for changes, while Product Groups are more concerned with the technical possibilities for combining and exchanging components.

If we evaluate this method of identifying product variants using the hourglass presented in Section 10.3.2, we see that only the information in the lower part of the hourglass is controlled. Each subsystem has a specification code and a standard manufacturing bill-of-material which lists the components from which this subsystem is assembled. Specifying and identifying product variants in this way can lead to the following problems:

- no support in assembling the finished product;
- no shared view on the boundaries of a sybsystem product family (i.e. the dotted line in Figure 10.8), indicating which subsystem variants can be combined;
- the need for developing ad hoc interfaces for individual product variants;
- difficulties in translating customer requirements into subsystems and components.

From the above, we can conclude that the classification of "engineer-to-order" is partly true for this approach to identifying product variants. Standard components and subassemblies are needed together with one-of-a-kind interfaces to create specific finished products.

Further, the lack of support during the final assembly process serves to freeze the position of the customer order decoupling point. The need to assemble and manufacture more products to customer order (e.g., components which are now manufactured based upon a pro-

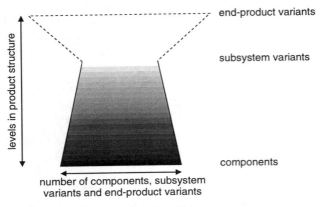

Fig. 10.8 Hourglass for identifying components

duction plan) is hindered by the lack of control over customer order dependent manufacturing.

10.5.2 Identifying preferred systems

In response to this engineer-to-order situation, management developed a new policy in which a limited set of finished products has been identified and published in a separate (second) catalogue. Since only a limited number of products are involved, they are manufactured based upon a production plan. As a result, this approach has some characteristics which resemble "make-to-stock" manufacturing. Bills-of-material are compiled for the finished products at the top part of the hourglass to support the final assembly process. This also enables an easy installation at the customer site since it is then possible to test all of the listed combinations of components beforehand.

The main drawback of this approach, however, is that it is does not satisfy the needs of the sales engineers and customers. The limited set of finished products is too restrictive for the customer. At the same time, only a few of these finished products make use of almost all subsystems and components, thus, failing to simplify the control and manufacturing effort of the Product Groups of Component Operations (Figure 10.9).

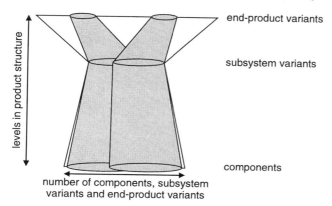

Fig. 10.9 Hourglass for identifying a limited set of finished products

10.6 FURTHER ANALYSIS OF THE PROBLEM ISSUES

Although the two ways of specifying product variants appear rather different, they nevertheless have some aspects in common.

- Perhaps the most important similarity in these two approaches is that only one view of the product family is supported, namely, a technical view describing either the components or the finished products in technical terms. This introduces the risk of a salesman or customer not recognising the economic aspects when he/she does not have the skills to translate a technical description into economic terms.

- Another important similarity is the difficulty in achieving agreement on the boundaries of a product family. There is often too much room for interpretation in both instances, regardless of whether a product family is specified by listing the components or by listing a limited set of finished products.

- A third similarity concerns difficulties in maintaining consistent product data, particularly when engineering changes occur. We already described in Section 10.5.1 that Commercial and Component Operations have different views on the rules which dictate how and when the component specifications need to be updated. A similar problem exists with respect to the specification of finished products.

It is clear that an approach for specifying product variants should address, minimally, the above-mentioned problems before it can be considered to be suitable for use in an assemble-to-order manufacturing environment. An approach which satisfies this requirement was developed at Eindhoven University of Technology: the generic bills-of-material approach. The Medicom case presented here is used again in Chapter 16 to illustrate how this approach can be used. Chapter 20 discusses the proposed production control system.

10.7 SUMMARY

The Medicom case is an example of a company situation in which systems can be assembled to customer order, but problems are encountered due to the methods used to describe product families and their variants. Medicom presently uses two radically different approaches for specifying its products: one approach describing the components which make up a family and the other approach describing a limited set of finished products. Neither of these methods adequately support an assemble-to-order manufacturing situation.

REFERENCES

1. Erens, F.J., *The synthesis of variety*, Doctoral dissertation, Eindhoven University of Technology, 1996

11

Make-to-order in work-flow-oriented manufacturing

11.1 INTRODUCTION

This chapter describes the case of customer driven production in the work-flow-oriented industry. More particularly, the factory described here produces sheets of fine paper in different quality levels, weights and packaging specifications, and in almost any size the customer might wish. The factory is of a **make-to-order type** because the size of the sheets and the packaging requirements are customer-specific.

11.2 COMPANY DESCRIPTION AND SITUATION

General characteristics

The company selected for this case study produces paper and solid boards, the factory concerned being in the fine-paper division. This division comprises several factories producing wood-free coated paper for high quality prints, e.g. art and other books, brochures and annual reports. Each factory has its own range of products. In this case study we are focusing on one of these factories, the one in question manufacturing paper sheets (instead of reels).

The annual turnover of the fine-paper division is about 400 million ECUs; the total number of people employed is 1,900.

The range of products of the factory

The range of products manufactured by this factory can be characterised by various parameters (figure 11.1). The first two parameters concern the paper as a material: the quality and the weight per square metre or *basis weight* (gr/m^2). We can distinguish four main qualities: brilliant, satin, matt and embossed. Brilliant is glazed paper often used for purposes such as posters, brochures, etc. This is not the case for the other qualities. Matt is non-glazed paper which can be used for writing, for example. Satin is an intermediate form between brilliant and matt. Embossed is matt paper which has been prepared to obtain a certain relief on the paper.

The basis weights made are between 115 gr/m^2 and 300 gr/m^2. Only the most common basis weights are illustrated in figure 11.1. Apart from these paper characteristics, the final product shipped to the customer is specified by:

format (length and width)
the fact of whether a customer wants the paper to be guillotine-trimmed (for clean edges on the pile of paper)
the method of packaging, and
the method of palletising.

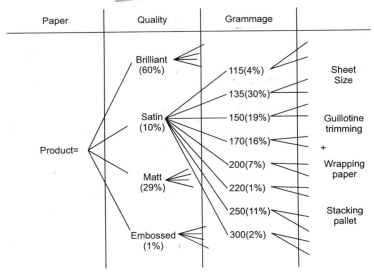

Fig. 11.1 Characteristics of the product

Sales market

The sales market of the fine paper division is concentrated in southern and western Europe (approx. 85%). Its customers are wholesalers and printers. Even when the paper is shipped direct to a printer, the wholesaler often acts as an intermediary. About 75% of total sales involve long-term contracts with wholesalers. These annual contracts are not worked out in detail but give only an indication of demand and the price agreements. Therefore, individual orders are still highly specific.

11.3 INTERNAL ORGANISATION (RESOURCES VIEW)

The internal organisation of the fine paper division is illustrated in figure 11.2. There are four areas of management: Human Resource, Production, Sales and Accounting. The production area consists of two production plants supported by several staff departments: Plant Planning, Quality and Product Development.

Each of the plants has a functional organisation (figure 11.3) which is based on the production process that will be described in the next section. In organisational terms, there are different production departments for the paper machine, the coater and finishing.

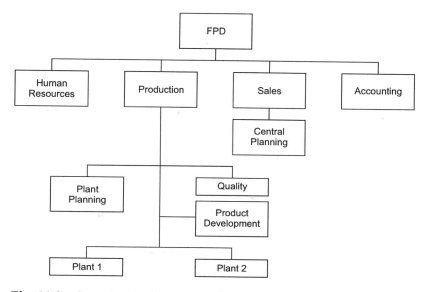

Fig. 11.2 Organisational structure of FPD

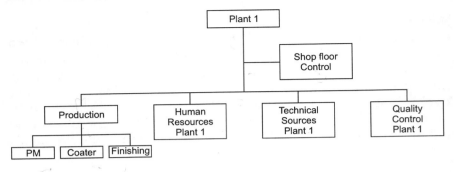

Fig. 11.3 Organisational structure of plant 1

11.4 THE TRANSFORMATION PROCESS (WORKFLOW VIEW)

The factory described in this chapter manufactures to customer order, which is an exception in papermaking. The reason for this is that fine paper is the top segment of the paper market. The physical transformation process is quite similar to other paper-making plants. However, the customer order entry process is different from these other plants. We shall first describe the procedures which are used for order entry in section 11.4.1, before describing the physical transformation process in section 11.4.2.

11.4.1 The Order Entry Process

Customer order entry takes place in the sales department of the fine paper division as a whole (see figure 11.2). The procedure is illustrated in Fig. 11.4. Suppose that — after negotiation — a customer has decided to place an order. The first action taken by the sales department is a market check (each market gets a fixed percentage of the total volume of production of each plant). If the market percentage is not exceeded, the order is booked into the information system. Next, a rough capacity check on total load can be made against the current Aggregate Production Plan. This capacity check is performed within the sales department by central planning. It is specifically concerned with the load on the most important machine: the paper machine or PM for short (see figure 11.3). The future capacity of this machine has been split up into production cycles of 10 days. Each cycle has a changeover plan for different paper qualities and basis weights, as will be explained in the next section.

When the total load booked for a certain run (i.e. production cycle) exceeds the capacity available on the paper machine, there is feedback to Sales. Central Planning (which is part of the sales department) acts as the intermediary here between Sales and Production. Before the order is downloaded to Production it can still be changed by the responsible salesperson in negotiation with the customer. Once it has been downloaded to Production (i.e. to a particular Plant Planning group), this is no longer possible.

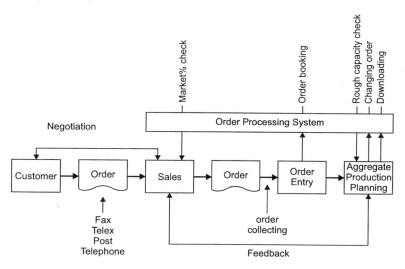

Fig. 11.4 Order entry

11.4.2 The production process

The primary process is illustrated in figure 11.5. Only the most important production steps are represented. The first step in the process is the paper machine (PM). This is an extremely expensive machine that produces paper from pulp in a continuous dehydration process. This will not be elaborated in greater detail.

The paper machine produces paper rolled up on a large iron axle or tambour. The axle plus paper is called a parent reel. These parent reels are more than 4 metres wide and have a diameter of about 2 metres. The weight of the parent reel at this point in the process is about 12 tons. The parent reels are first transported to a rewinder, where the edges are trimmed and any production errors remedied.

The parent reels are then transported to the coaters. Either one layer of coating (lower basis weights) or two layers (higher basis weights) are applied to both sides of the paper. The kind of coating used determines the quality of the paper (brilliant, satin, matt or embossed). After being coated, the paper is again wound onto a parent reel.

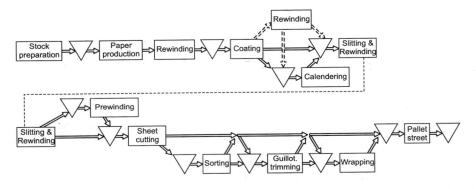

Fig. 11.5 Production process

The next step in the process is calendering: by leading the paper through a number of iron rollers, the paper acquires a smooth, shiny surface. Calendering is only required for brilliant paper. Matt, embossed and satin papers do not need to be calendered.

After calendering, the main stream of orders requires that parent reels first be slit (cut in width) by slitting knives (figure 11.6) and rewound onto smaller reels, before these smaller reels are cut into sheets and packaged. This results in the following main processing steps:

- slitting and rewinding (fig. 11.6)
- sheet cutting (fig. 11.7), and
- palletising.

Apart from these main steps, several minor steps are possible. These are shown in figure 11.5.

As figure 11.6 shows, the large parent reels are first slit and rewound onto smaller reels. The paper is unrolled, slit and wrapped on a cardboard cylinder. Next, at the sheet-cutting machines, these smaller reels are placed in a mill and cut into sheets. Depending on the basis weight of the paper, a fixed number of reels can be cut at the same time (figure 11.7). Sometimes some preworking (prewinding) is carried out in order to achieve maximum loading at the mill.

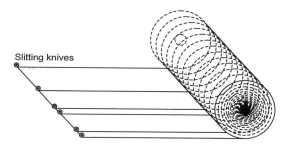

Fig. 11.6 Parent reel (tambour) being slit

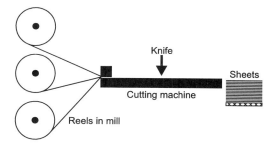

Fig. 11.7 Sheet cutting

Once the paper is in the form of sheets, it has to be packaged. In some cases, these sheets first have to be sorted, or have to be guillotine-trimmed or wrapped in paper (see figure 11.5). About 45% of the orders have to be packed in wrapping paper before being palletised, the others can be palletised without being wrapped. The paper that is not packaged in boxes or in paper is wrapped around with plastic. The larger part of these orders is automatically wrapped at one of the packaging machines, a small amount being wrapped by hand. However, all orders pass the pallet street in one way or another. In this street, pallets loaded with paper are weighted, have a ribbon affixed and are labelled. After palletising, the order is ready for shipment.

The course of an order deviates from the main streams when preworking is required before sheet cutting or when, after sheet cutting, the paper has to be sorted or guillotine-trimmed. Apart from the operations mentioned here, a number of additional operations are possible in finishing the orders. These additional operations are mostly due to production errors, but some may also be a customer requirement, e.g. guillotine-trimming.

11.4.3 Space Requirements

During the first stages of production the paper is rolled onto large iron axles (tambours). These reels are slit and rewound onto smaller reels and finally cut into sheets. Since the size of the objects diminishes with each step, more space is required at the end of the process than at the start. For instance, about twice as much space is required to store an equal amount of paper on reels than on tambours. The amount of space available in this part of the building is limited. In front of the sheet-cutter there is room for about 18 hours of paper production, in front of the packaging machines there is even less. When there is not enough space in front of the sheet cutters, reels are transported to a higher floor. It goes without saying that this kind of material handling is rather expensive. Therefore, production planning should aim at avoiding situations where material in process requires more space than is available.

11.5 THE PRODUCTION PLANNING SYSTEM

In order to describe the production planning system, goods flow control (GFC) at factory level is distinguished from production unit control (PUC) at shop-floor level [Bertrand et al, 1990]. Remember that the organisational structure of the factory shows three departments, viz. Paper Machine, Coater, and Finishing (see figure 11.3). Despite the fact that these organisational structures do exist, only one production unit is distinguished from a logistics point of view. In other words, the whole factory is considered to be one production phase without intermediate stocking points or work-order-release points. An outline of the production planning system is illustrated in figure 11.8.

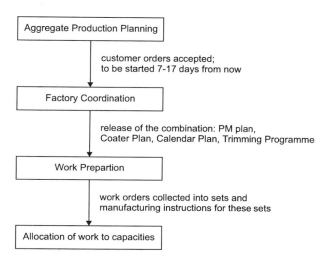

Fig 11.8 Outline of production planning

11.5.1 Goods flow control at factory level

At factory level it is possible to distinguish two decision functions: aggregate production planning and factory coordination.

Aggregate production planning
In this factory, aggregate production planning is only concerned with the paper machine. The starting-point of control is a production cycle based on a fixed changeover sequence on the paper machine. The production cycle is assumed to have a fixed length of 10 days. The paper machine produces a fixed number of tons of paper per day. Orders are booked in a cycle until the total amount of tons booked exceeds total capacity. Each basis weight can be produced twice in this cycle (see figure 11.9.). The choice of the changeover sequence is based on the minimisation of changeover losses on the paper machine. The length of the cycle is bounded by the lead times

acceptable to customers and by problems that occur on the shop floor if the paper machine were to produce the same quality for a long time. In order to save changeover time, the cycle starts with 115 gr/m^2, climbs up to 300 gr/m^2 and drops again to 115 gr/m^2. The length of the cycle is 10 days, which means that the normal lead time for customers varies between 3 and 13 days (if the activities in the Finishing Department take 3 days).

As a rule, the brilliant and satin are produced in the first part of the cycle and the second part is reserved for matt paper. However, departures from this rule are common in order to avoid problems on the shop floor. These problems, which may occur in particular at the calenders and at the slitter-rewinders, will be described in detail under the next heading. Because of these problems, there is a general rule that the cyclic production plan should not exceed 10 days, and there is a common understanding that matt and brilliant parts of the cyclic plan may be interchanged sometimes at the request of the production department.

The cyclic aggregate plan has a three-month horizon and is under the control of Central Planning in the sales department of the Fine Paper Division. The plan contains the expected starting date for each cycle and shows for each cycle how much workload (in tons) has already been booked and how much is still open. These values are needed for a rough capacity check. Note that Central Planning checks only the total number of tons booked in a 10-day cycle. More detailed plans are left to Factory Coordination.

Factory coordination (The so-called PM plan and trimming programme)

The customer orders booked are downloaded from Sales (Central Planning) to Production seven days before the start of a cycle on the paper machine. Within Production, the Plant Planning function (see

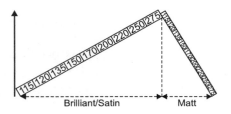

Fig. 11.9 Cyclic production plan

figure 11.2) combines the orders with identical quality and basis weight properties into a so-called "combination". Plant Planning makes a short-term plan for production on the paper machine containing detailed information on the production of a combination (time, quantity, individual customer orders). This so-called PM plan is comparable to the Master Production Schedule in other factories. It is the basis for all other planning activities at factory level. However, it serves at the same time as a work-order-release plan for the factory. More specifically, the PM plan drives similar plans for the coater, the calenders and the slitter-rewinders. These plans are made by Plant Planning together with the PM plan. This is necessary because the results of the two plans last-mentioned can influence PM planning, as we will see.

The plan for the coater follows straightforwardly from the PM plan by keeping the same planned sequence of orders. The plan for the calenders is less straightforward. It should be borne in mind that only brilliant paper is calendered, but that the calenders together produce at 75% of the speed of the paper machine. Therefore, when brilliant paper is produced the material will stock up on parent reels in front of the calenders. This results in additional material handling with the associated quality problems and waste of human time. Also, a shortage of tambours may occur, which would result in a forced shutdown of the paper machine. These problems must be avoided by every means. Therefore, contrary to figure 11.9, matt is sometimes interspersed with brilliant in the PM plan.

The plan for the slitter-rewinders (the so-called trimming programme) is rather complicated. First of all, the customer orders of a particular combination have to be grouped into several letters. These letters are formatted at the slitter-rewinder (figure 11.10). A set is characterised by its width and its number of reels (representing length). In the second step, these sets are mapped on the two-dimensional surface of a combination. This step is supported by a program (the so-called trimming programme), which is based on a class of algorithms well-known in operational research (two-dimensional cutting stock algorithms). After the second step, it is clear how the combination will be slit. It is necessary to draw up this plan at Factory Coordination level because it also specifies the expected loss of material in trimming and therefore the exact amount of paper to be produced for a particular combination. Consequently, the trimming programme influences the PM plan.

130 *Make-to-Order in Work-Flow-Oriented Manufacturing*

Finally, the PM plan is used for issuing the materials required in the production process (pulp, ingredients, coatings, wrapping paper). These materials are always available in the factory, but inventory control of them is separated from production planning.

The above discussion shows that Factory Coordination considerations may interfere with Aggregate Production Planning. Consequently, there is a feedback loop from Factory Coordination to Aggregate Production Planning (or in organisational terms from the Production Department to the Sales Department).

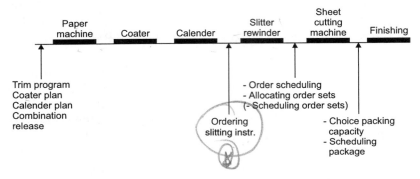

Fig 11.10 Grouping the orders of a combination into letters

11.5.2 Shop-floor control within the production unit

After release of the combination, several decisions have to be made. In this section we describe which decisions are made, when they are made and what decision-making rules are employed. Two kinds of activity can be distinguished: work preparation and allocation (see figure 11.8).

Work preparation involves first of all order collecting. This is an activity which collects sets of orders suitable for being processed together at a particular manufacturing stage. Next, work preparation involves issuing detailed instructions to the shop floor. So, work preparation tries to optimise machine efficiency and quality.

Allocation allocates work to capacities and determines priorities in queues. Thus, allocation and scheduling try to optimise human efficiency and due-date reliability.

Work preparation

As stated, work preparation starts with the collection of sets of orders which are to be processed together. However, the highest level of order collecting (the combination of orders with the same quality and basis weight) is performed at Factory Coordination level when a combination is released, as has been discussed already. A combination is closed (or released for production) several hours before production commences. At this point it is no longer possible to download any orders for this combination.

A second level of collecting sets of orders to be processed together is also performed at Factory Coordination level, viz. the concept of letters resulting from the trimming programme. Here, too, a letter can be seen as a collection of orders with common properties (actually, the width of the paper).

Work preparation can now start generating the machine instructions for the PM, the coater, the calenders and the slitter-rewinders. The PM, the coater and the calenders can be prepared straightforwardly because these machines produce parent reels. The slitter-rewinders require more attention because instructions need to be generated on how to cut parent reels into smaller reels. Generating the manufacturing instructions is based on the trimming programme. Each instruction for the slitter-rewinder shows the required positions of the knifes and shows how many tambours should be slit and rewound in order to obtain the required number of reels for each set, and the correct widths. The cutting speed, based on the number of parallel cuts and the basis weight, needs to be determined. In general, the speed increases when the basis weight goes up and decreases when the number of cuts goes up.

Work preparation at later stages in the finishing phase may also start by collecting sets of orders. This holds in particular at the sheet-cutting machines. The capacity of the sheet-cutters is determined by the number of reels that can be hung in the mills (see figure 11.7). This number is higher when the basis weight is lower. (Notice that this situation is opposite to the one at the slitter-rewinders). However, it is very difficult to predict which reels will be produced by the various slitter-rewinders at which point in time. Because the space in front of the sheet cutters is very limited, order collecting is carried out by the foreman and is combined with allocation and scheduling. This will therefore be discussed below.

Allocation

Paper machine, coater and calenders. We have already discussed the fact that the cyclic production programme (see figure 11.9) is used for scheduling the PM, the coater and the calenders. After release of a combination, the foreman on the shop floor allocates work orders to machines. In principle, these decisions follow the plans made in Factory Coordination. However, due to capacity problems and various types of disruption, rescheduling is often necessary. Guidelines for allocation and scheduling are available at each of the machines. Very often, though, decisions are made ad hoc.

Slitter-rewinders. In general, the principle of first-come-first-served (FCFS) is applied at the slitter-rewinder. No distinction is made between the different parent reels belonging to a certain combination. The slitter-rewinders have a slight excess capacity. After the slitter-rewinder, the tambour irons return to the paper machine. Since the excess capacity is slight and the number of tambour irons limited, it is generally accepted in the factory that the slitter-rewinders should always run. Higher basis weights especially will be given priority because the operating speed (in tons/hour) for these products is higher. Priority is also given to the higher basis weights of matt paper for quality reasons.

Finally, the priorities at the slitter-rewinder are influenced by changeover costs, maintenance problems, the characteristics of human operators and the expected workload at the sheet cutters. For all these reasons, the final priorities are set by the foreman on the shop floor.

Sheet-cutting machines. After slitting and rewinding, the order sets are allocated to the sheet-cutting machines and the sequence of the orders within an order set is determined. The sequence of the order set at each of the sheet-cutting machines is again FCFS. Within one order set no distinction is made between the reels. Although each type of reel has a favourite machine, there are no special rules for allocation.

Allocation is made on the shop floor by a foreman. If the amount of space in front of the sheet-cutter is insufficient, priority is often given to order sets containing large orders with few changeovers. Occasionally, priority is also given to orders with specific packaging properties when the workload at the packaging machines is too low or too high. These priority rules only apply to the orders within an order set.

Packaging. After sheet cutting, the final operations take place. Two decisions which still have to be made are the choice of packaging machine and scheduling at each of the packaging machines. Just like elsewhere in the finishing department most of the priority rules are capacity-based. In all circumstances rush orders have highest priority in these final operations.

11.6 FURTHER ANALYSIS OF THE PROBLEM ISSUES

Further analysis reveals a number of important problem areas disrupting the flow of goods in the factory. We can distinguish four problem areas that are interrelated: capacity, human resources, product quality and lead times. These problem areas will be discussed below.

11.6.1 Capacity problems

In an ideal situation the paper machine and coating machine produce paper, with finishing activities following this process according to flow-production principles. However, the situation is not ideal. The calenders, the slitter-rewinders and the packaging machines occasionally have to deal with capacity problems, as discussed above. For the sheet-cutting machines these problems are structural.

The sheet-cutting machines. The problems at the sheet-cutting machines are more structural. The capacity of the sheet-cutter is insufficient to work up the output of the slitter-rewinders. As a result, reels frequently have to be moved to other factories. Furthermore, there is not enough space in front of the sheet cutting machines, so that reels have to be moved to an upper floor. Because of these space problems the planner/preworker obviously tends to give priority to orders that are running smoothly. The difficult orders are moved upstairs and eventually become rush orders.

The capacity requirements of particular types of paper at the slitter-rewinder and at the sheet-cutting machines are the opposite of each other. High basis weights are produced efficiently at the slitter-rewinders but these are inefficient at the sheet cutters. The opposite is true for low basis weights. When the high basis weights are produced at the slitter-rewinders, the shortage of space and lack of capacity on the sheet-cutting machines is felt most significantly. With low basis weights, the sheet cutters usually starve due to lack of work being supplied. The fact that work at the sheet cutters sometimes has to be subcontracted is perceived as a problem, especially when substantial idle times occur shortly before the need for subcontracting arises.

11.6.2 Human Resource problems

Each employee has a fixed task. In a factory with fluctuating machine capacity this is not an ideal situation since it is not possible for one person to change to another operation. Furthermore, it is often seen as a downgrading of the job when, for instance, someone at the sheet-cutting machine is also in charge of internal transport. It will be obvious that this lack of flexibility is harmful to lead times.

Another important problem is the authorisation of decision making. A lot of the decisions are made at the morning meetings and during daytime when the authorised people are present. At night a large number of the decisions cannot be taken because the authorised person is not present. For instance, when there is a quality problem at one of the tambours, the tambour will stay where it is until after the morning meeting. This, of course, results in a delay in the goods flow.

11.6.3 Quality problems

Quality problems occur in every factory and papermaking is no exception. Although the total amount of paper to be scrapped is only a few per cent, there is at least one problem at each stage of production during almost every shift. These problems have to be discussed by daytime staff who are authorised to prescribe corrective action. However, this method of working not only causes much delay for the reel or pallet with a problem, but also for the combination or set to which this reel or pallet belongs. This stems from the fact that such combinations or sets have to be processed as a whole. Consequently, quality problems frequently disrupt planning. The more "downstream" stages of production are more susceptible to this effect. Consequently, no plan is made for the sheet cutters and subsequent machines, and their loading is left to the foreman.

11.6.4 Lead times

Finally, the company faces a problem with its customer lead times. In the past, it was acceptable to quote customer lead times of at least 20 days. This time is needed with the system described above, due to the 10-day production cycle which must be downloaded 7 days before the start. Moreover, production lead times also require at least three days. However, time-based competition requires response times that are much faster. The company loses orders simply because customers need fine paper within one week. How can this be achieved? This question will be addressed in Chapter 19, which describes production planning for customer driven manufacturing in process-oriented companies.

11.7 CONCLUSION

This case study shows customer driven manufacturing in a process industry, i.e. a work-flow-oriented industry. Production management is difficult here, because many constraints play a role that interfere with planning decisions. These constraints may take the form of limited machine capacity, limited storage capacity, limited human skills, tight due dates, and complications in case of quality problems. Despite the flow nature of the production system, there are considerable short term variations in work-supply to several work centres.

The direct cause of the variations in supply at some of the machines is the use of a production cycle which starts with the lower basis weights, goes up to the higher ones and then back again to the lower basis weights (see figure 11.9). Due to quality and capacity reasons this production cycle cannot be changed. Good planning policy could provide a solution to the capacity problems, but there are problems with planning.

Poor order progress control

Lack of proper recording facilities makes it difficult to verify order progress, especially in the finishing department. During the day people are walking through the factory looking for orders, reels and sheets. The only means of progress control used is a rush-order list, which contains all orders to be finished within four days.

Local control
Each machine is controlled as a single unit, and many decisions are taken ad hoc. The result of this is that production programmes and operation schedules are changed frequently. Added to this, there is no understanding of the impact of such changes on the rest of the process.

Creation of internal rush orders
Using the rush-order list creates other internal rush orders. A large rush-order list disrupts the internal goods flow. Orders that are moved to the upper floor often appear on this rush-order list at a later stage, and they will only be finished after they appear on this list.

Unforeseen disruptions
Unforeseen disruptions are usually due to three kinds of deeper cause: human error, the complexity of the paper-manufacturing process and the absence of people or materials.

Human errors nearly always result in disruptions in planning, but the simplest situation is where errors lead to poor paper quality. This may result in scrap, meaning that the work either has to be redone (such as in the case of scrap at the PM) or that less paper or a different paper than required is produced. Each of these situations disrupts planning.

The complexity of papermaking often means that quality problems cannot be solved immediately, which is another cause of disruption in planning.

REFERENCES

1. Bertrand, J.W.M., J.C. Wortmann and J. Wijngaard, *Production Control. A Structural and Design Oriented Approach*, Elsevier, Amsterdam, 1990.

12

Engineer-to-order in resource-oriented manufacturing

12.1 INTRODUCTION AND DESCRIPTION OF THE COMPANY

12.1.1 General characteristics

The company selected for this case produces ships and other maritime products. It is part of a technology-oriented holding whose core business is maritime technology. This holding is divided into five divisions (see figure 12.1). The company is part of the shipbuilding division that has six shipyards, each with its own range of ship classes.

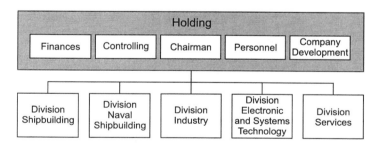

Fig. 12.1 The range of products

Because the operations are resource oriented, the shipyard offers a wide variety of maritime products such as:

- Container ships,
- Frigates,
- Roll-on/roll-off ships,
- Passenger liners,
- Corvettes,
- Multi-purpose cargo vessels,
- Reefer ships,
- Supply ships,
- Research vessels,
- Passenger-cargo ships.

Because these products are unique and ambitious technical objects, all of these products contain numerous and varying technical systems (see figure 12.2). This large product range can only be supported by customer-order processing that is consistently engineer-to-order oriented.

Each ship is individually designed and constructed according to customer specifications. The shipyard's fundamental decision to follow a distinct differentiation strategy requires permanent monitoring of the capabilities of employees and machinery. This resource orientation is restricted by emphasis on developing a pioneering design concept for the layout and construction of container vessels. This concept considerably improves reliability at a higher dead-weight and container intake while simultaneously lowering costs, and offers numerous advantages in terms of assembly methods. Because of this investment, the shipyard's production is not only resource-oriented but also partially product-oriented. This development activity was launched based on the fact that the yard delivered five container ships only in 1992.

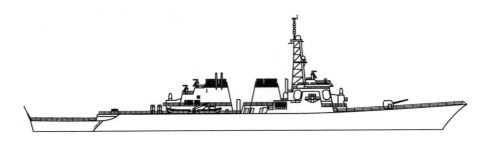

- Steel Hull Structure
- Accomodations
- Systems for
    - Energy Generation
    - Energy Transduction
    - Energy Distribution
- Equipment for
    - Cargo Handling
    - Transportation
    - Shops
- Safety Devices

- Systems to distribute
    - Sea and Fresh Water
    - Steam and Condensate
    - Air
    - Fuel and Diesel Oil
    - Lubrication Oil
    - Hydraulic Oil
- Systems for
    - Information Transmission
    - Communication
    - Monitoring and Control

Fig. 12.2 Ships - Multisystem Products

12.2 THE INTERNAL ORGANIZATION (RESOURCE VIEW)

The internal organisation of the shipyard is divided into three main areas: Sales, Manufacturing and Administration. The Sales area is responsible for preparing bids. This preparation work includes the technical specifications, cost calculations and scheduling the project for quotation of the delivery time. These activities require highly-qualified human resources with considerable experience in shipbuilding and detailed knowledge of both the resources and the shipyard.

The Manufacturing area has three main departments: design, operations planning and production. The design department is subdivided into the main groups "steel hull design" and "outfitting design". The operations planning department also has two work preparation groups for "steel hull" and "outfitting". The production planning groups are part of the operations planning department. This functional structure reflects the production process described in the next section.

The physical production process is highly resource-oriented. As will be seen in the next section, the availability of a large crane, for example, determines the nature of work preparation. The next section will also show that the workers in the outfitting areas are autonomously authorised to determine the best way to fit the piping, and are capable of doing so. This also reflects resource orientation. The major investment in resources, however, is probably the WOST concept, which is briefly introduced here.

Nowadays most shipyards throughout the world perform outfitting work on site, on board of the vessel. As a means for distinguishing itself, the shipyard in this case study implemented what it calls the "Workshop-Oriented Ship Production Technology" (WOST) concept, which resulted in significant shortcuts and reductions in outfitting costs.

The shipyard has invested in an enormous hall where indoor outfitting is performed. Outfitting-intensive modules, such as the engine room module, are transferred to the outfitting hall. Here the outfitting is carried out in a controlled environment and walking distances for workers are reduced to a minimum. After lifting the pipes into the various stations (decks) of the engine-room block, transport and assembly are carried out manually. Upon completion, fully-equipped modules weighing up to 3,800 tons are transferred directly to the building dock with a heavy-duty transport system.

A similar process was adopted for the structurally-complete steel deckhouse. Open galleries lead from the WOST centre to the various levels of the deckhouse. Material and equipment flow directly from the storerooms and workshops, resulting in rapid and efficient outfitting. This is clearly an investment in resources that has a major impact on the work-flow through the yard and on many other design choices and performance indicators.

12.3 THE TRANSFORMATION PROCESS (WORK-FLOW VIEW)

Due to the short delivery times required by the customers, there is significant overlapping between the various order-throughput phases as compared to the situation in repetitive production. Because of this overlapping and customer influence through to final assembly, the working instructions are often compiled by the shop-floor employee without involvement of either the designer or the planning expert. The manufacturing phases and their main activities as indicated in figure 12.3 are discussed below.

Once an owner has identified the need for a new ship and defined the ship's operational requirements, the shipyard's first activities involve a preliminary definition of the vessel's basic characteristics. Depending on the customer requirements - a shipping line, for example, may anticipate the need to annually transport 120,000 automobiles between Japan and Northern Europe - a preliminary definition of the ship must be specified, including main dimensions, hull shape, general arrangement, powering, machinery arrangement, mission systems definition, capacities for variable weights (such as fuel, water and stores) and a preliminary definition of the major systems (such as the structural, piping and electrical systems).

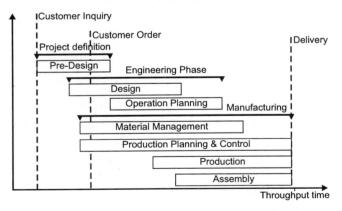

Fig. 12.3 Project definition, Engineering and Manufacturing phases

The various design methods currently applied in ship design are often based on a trial and error process employing an increasing degree of refinement; accuracy and a higher level of detail characterises the basic approach to ship design. The well-known design spiral (figure 12.4) illustrates the basic principles of this widely-accepted approach to ship design (see Hassan and Thoben [1992]), which differs from the general methodologies for mechanical products (cf. VDI [1993]).

Because more and more quotations have to be made under increasing time pressure - with a success rate of less than 10% - more attention must be paid to quotation planning. On the one hand this involves computer-supported product cost and price calculation. On the other hand, quotation planning must render an early estimation of production time and capacity demands, and the relation to available resources and capacities. If an offer becomes an order, all subsequent planning activities are based on these results.

As a result, the main objective is early estimation of a realistic delivery date and the latest date production can be started, taking

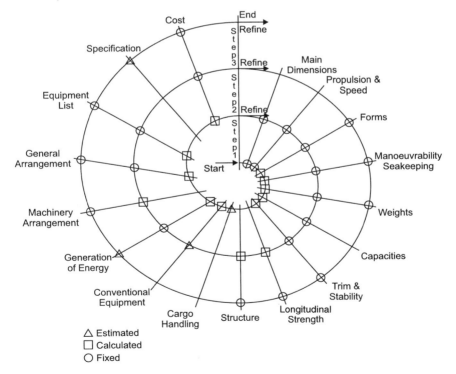

Fig. 12.4 Pre-Design Spiral

alternatives into account such as the use of external facilities and the evolution of a multi-organisational project.

The success or failure of a quotation and therefore the subsequent order primarily depends on the skill and knowledge of the decision-making employees, whose experience compensates for the lack of information concerning a customer's request at this early stage. Almost all part weights related to the steel construction, for example, are estimated based on the steel construction engineer's experience with a precision as high as 98% (according to statements from several staff members).

12.3.1 Engineering phase

The design process can be divided into two general categories of work:

- the steel-hull and steel-structure design (steel sections), and
- the design of outfitting parts (machinery, accommodations, pipe systems, etc.).

Steel-hull and Steel-structure Design
Based on the results of the pre-design (e.g. geometry of the steel hull) a draft of the steel construction is prepared prior to the detail-design and calculation phase.
The initial layout generates the following important documents containing planning parameters of the detail design:

- classification drawings;
- general arrangement plan;
- steel scantling plan with section plan;
- bills of materials steel (profile list, plate list).

The unique geometry of most of the steel profiles and plates for each ship stresses the one-of-a-kind nature of shipbuilding. The preconditions for the detail design phase are determined by the results of experiments involving the form of ship lines (already determined during the pre-design phase), hydrostatic and hydrodynamic analyses, classification requirements, laws and rules.

In addition to hydrostatics and hydrodynamics, until recently the maximum lift of the frame crane was the weight determination planning aspect. A heavy-duty transport system has now been installed which overcomes the restriction; the weight capacity has been increased from 450 tons to 3,800 tons. Now the most difficult part of planning and executing the steel design is the availability of information and product data (CAD-data) from the outfitting design department.

The availability of this information is determined by the yard's design priority rule: the steel construction has to be made first, followed by the outfit construction. For example, the steel designer has no exact description of or data concerning materials such as pipes, which must be coordinated with the outfit design. If the specifications for pipe holes were known earlier, these holes could be mechanically cut into the steel profiles, which is less expensive. If this information is unavailable, the cutting or welding (hot works) must be carried out 'on board', unscheduled, resulting in expensive operations.

The drawing process generates workshop drawings, a total drawing plan, control part lists, a box plan, documents for the flame-cutting and profile-bending machines, etc. Because the shipyard's design capacity is insufficient to finish all construction activities on time, between 10 and 90 percent of construction and drawing time is subcontracted to external engineering companies or other yards. Inquiries are necessary when unscheduled capacity changes occur during the design process. Because the licence policy of the CAD-system vendor facilitates subcontracting and because the market also offers enough suitable engineering capacity, engineering companies can easily work with CAD-licences offered by the yard.

Outfitting Design
Outfitting design includes all typical two- and three-dimensional non-free-shape mechanical design activities. In addition to the design of the machinery, the design of the various piping systems for water, fuel, etc. is one of the most intensive outfitting design activities. Based on the pipe systems, the yard distinguishes between pre-fabricated and schematic pipes.

The location of aggregates such as engines, pumps etc. is essential in determining the specifications of the pre-fabricated pipe systems and the engine-room layout plan. These systems are the basis for constructing a 1:10 or 1:20 geometrical model of the engine room. The designer takes the positioning information and the required dimensions from this model and transfers this information to an isometric drawing. The relevant manually-generated product data (isometric drawing, NC code, etc.) is the basic document for production and production planning.

For the schematic pipes, however, there are no detailed drawings or bills of materials. The only type of drawing for these kinds of pipes are schematic drawings describing the schematic connection of the related pipe systems. Schematic pipes have a smaller diameter and are designed 'on board', taking the topology into account. A manually-bent wire model is constructed based on the actual geometry for this purpose. The pipes are either straight, bent or welded together.

Operations planning

The responsibilities of operations planning are defining operations and process plans, calculating operation standard times and assigning process plans to the defined BOMs. A major problem is created by the different views and structures of the BOMs. During the design phase, BOMs are organised according to the relevant engineering discipline, which is necessary in order to generate drawings. Because the manufacturing process cannot be organised according to engineering disciplines, the BOMs must be transferred to a manufacturing-oriented structure. This work is currently the responsibility of operations planning, but will be carried out by the design department in the future. The calculations for operation standard times are based on simple tables of standard times.

Unlike operations plans for mechanical products such as machine tools, the operations plans for piping and sheet-metal components contain only rough operation descriptions.

12.3.2 Manufacturing phase

The first manufacturing activities are initiated shortly after the first design activities. As described in figure 12.5, the manufacturing and assembly processes in the shipyard can be subdivided into different phases. Due to the growing size and weight of the various objects (from parts to panels and sections), there is a point during steel assembly (phase 2-3) at which a transition must be made from "moving product to process" to "moving process to product". Phases 1 to 3 can therefore be performed according to the workshop or line production principles as described in Chapter 15 (for example pipes, sheet metal, accommodations). Phases 4 to 6, the main shipbuilding processes, are performed on the construction sites.

The transformation process (work-flow view) 145

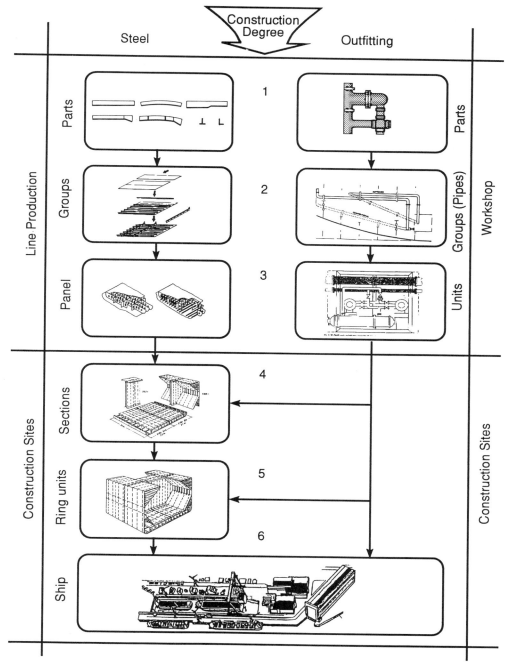

Fig. 12.5 Levels of work in manufacturing

Pipe manufacturing

The yard's pipework and outfitting facilities are located in the central part of the shipyard. The pipe fabrication shop (phases 1, 2, and 3 on the right-hand side of Figure 12.5) is responsible for receiving pipes from the pipe stock and for all the relevant fabrication steps such as cutting, bending, welding, assembling and testing the pipes and partially-mounted systems, and for the various sorting processes. The various conservation processes are also supervised by pipe workshop personnel. The last physical activity is staging palletised pipes and inserting these pallets into the right buffer. The pallets are issued for assembly in sections, in ring units or in the complete ship.

Steel Sections

Steel parts, small groups and panels (phases 1, 2, and 3 on the left-hand side of Figure 12.5) are fabricated in the steel workshop. The steel sections are assembled from these components on a construction site. Whenever possible, the outfitting work (installation of pipe systems in steel sections) is already started at the section level (phase 4 in Figure 12.5), which is a very early production stage. This gives the workers direct access to the compartments where the pipes are to be installed. The assembly is carried out according to shop-floor drawings of the relevant section.

Upon completion of the section assembly, the modules are coated and then buffered along the building berth (dry dock).

Dock assembly

While buffered alongside the building dock, the outfitting work of all modules can proceed so that nearly all steel units are at an advanced outfitting stage when they are lifted in the dock for the final hull assembly.

During the step-by-step final hull assembly - the junction of pipelines and systems with other pipelines and systems, components and ship-oriented structures - individually-adapted fitting pipes are usually needed to solve all geometrical tolerances between the modules. Due to the uniqueness of these pipes, the production differs from the sequence for pre-fabricated pipes. Before fabrication, a wire model consisting of pre-mounted flanges and connecting wires is constructed based on the local geometry on board. After this has been done, the entire model is dismounted and carried to the workshop, where the fitting pipes are fabricated as one-of-a-kind products.

Because the WOST concept has been implemented and based on the high outfitting level of all modules, the entire duration of the final hull assembly and the relevant dock occupation can be limited to 6-8 weeks for a container ship. This dock is the shipyard's critical manufacturing resource. After leaving the dry dock, final assembly of the vessel is completed at the outfitting quay. Before the ship is delivered to the customer, extensive on-board tests must be performed during a first run.

12.3.3 Inter-organisational production

In order to meet the agreed delivery time within the calculated budget, the shipyard must apply the simultaneous engineering concept during the entire manufacturing phase. In doing so, the yard must take into consideration that intra-organisational processes are insufficient. On the contrary: the required flexibility of resources and concentration on core activities (and thus the reduction of the manufacturing penetration) require close involvement of various types of suppliers (see Figure 12.6). These are engineering consultants, suppliers of systems such as main engines etc., raw-material suppliers, subcontractors and suppliers for consigned component production.

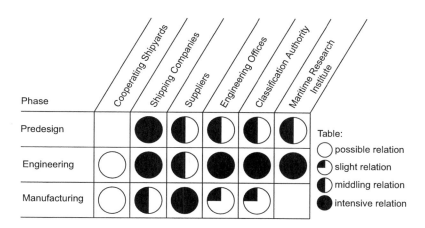

Fig. 12.6 Supplier involvement

148 *Engineer to order in resource-oriented manufacturing*

The design phase of the production process in particular is affected by the fact that classification societies must be involved with reference to ship classes. Three main steps can be distinguished that are required for each customer order before the required certificates are issued:

- Design approval preview.
  This is a rough check of the design to determine critical areas for thorough investigation and conformity checking. This preview is often performed during the bid preparation phase.
- Primary structure check.
  This consists of approval of the primary design as part of the certification process. This approval process can result in requests for design changes.
- Secondary structure check.
  Unlike the primary structure check, the approval areas involved in the secondary structure check are the ship hull, the machinery and all aspects concerning safety and pollution regulations agreed upon in international conventions. In addition, technical areas such as electrical systems and instrumentation, piping, bridge and automation, etc. are checked to the extent deemed necessary by rules. This check may also result in requests for design changes.

12.4 CURRENT PRODUCTION PLANNING AND CONTROL

At this time, the shipyard has implemented four production planning level concepts. The levels can be distinguished based on the planning horizons and degrees of data aggregation and accuracy.

12.4.1 Rough planning

The shipyard's rough planning consists of multi-project scheduling and capacity planning.

The capacity planning distinguishes between twenty production units, not including the design capacities. Not so much machine resources but spatial resources, e.g. production premises such as halls, are considered. The available capacity of a production unit is calculated based on the amount of relevant work, the average weekly working hours and the average percentage of missing hours.

The capacity demands for each production unit are specified only in relation to the various manufacturing phases of a project (and not to production activities). In order to achieve suitable time distribution of capacity demand, "type curves" are used. These type curves describe how the required capacity for a production unit in a particular phase is spread over time.

By comparing capacity demand and available capacity, periods of free capacity or overload capacity are determined and corresponding measures are taken.

The spatial resource planning is performed manually without any support from an information system. Because the project demand for spatial resources is defined with respect to production phases, this results in spatial resource planning based on the scheduling.

The rough scheduling is performed using hierarchical project network plans (extended Metra-Potential Method). According to this method, different levels of detail can be applied simultaneously. The number of network plan levels, up to a total of six, depends on the product and the customer's requirements. In order to reduce the effort required to define an order-specific network plan, former projects and parts of standard network plans are used.

On the lowest level of network plans (with the highest level of detail), activities are partly defined with durations of less than one day.

12.4.2 Design planning

The planning of drawing involves additional partners and controlling tasks. At the end of the drafting stage, the general arrangement plan must be sent to and confirmed by the classification society as defined in the contract. During drawing iteration, additional classification drawings such as for the steel plan, the midship section, bulkheads etc. must be delivered to the classification society. The time involved in document distribution (in-house and external) must also be planned and documented. The in-house activities, e.g. document output and arrival, and the time required by the classification society are planned in drawing time-sheets, including buffer times. These drawing time-sheets are often based on existing sheets for similar ships.

The worst case scenario - rejection of a classification drawing or calculation - must also be planned, which means that time and capacity buffers must exist based on which re-design and re-calculations can be performed.

150 *Engineer to order in resource-oriented manufacturing*

The planning and control of the design activities are based on the scheduling for the drawings. Every drawing produced is therefore listed by the design department.

12.4.3 Medium-term planning

The medium-term planning is based on the operations plan as generated and is performed with the help of a Production Planning and Control (PPC) system. The PPC system schedules the operations plan, taking the time dependencies and availability of materials into consideration, but does not check the availability of human resources and machines. The analysis of capacity loads must be performed manually. Evaluation sheets can be generated by the PPC system for selected resource units or groups and for a specified time span. Existing capacity overload situations must be solved by the planning experts. The alternatives are internal and external placing, modification of due dates and modification of scheduling parameters.

The medium-term planning is subdivided into steel manufacturing and outfitting and is performed centrally.

12.4.4 Manufacturing control/Short-term scheduling

In resource-oriented manufacturing, it is imperative that all due dates are met. This means that the scheduler (foreman) does not have the freedom to shift the due dates in the schedule, but can only manipulate the available resources. The shipyard's resources can be divided into fixed resources such as docks, cranes and floor space, and flexible resources such as small tools and manpower. On a larger construction site, the manpower has limited individual specialisation, meaning that to a certain extent the workers are freely allocatable within their profession.

The sequence in producing schedules is as follows: the due dates for the work to be done are determined, after which the content of the work to be done is determined; then the work is time-tabled for the rough resource groups. The final step in the procedure is the allocation of resources to the individual tasks. The process is essentially the same for all resource types.

12.5 FURTHER ANALYSIS OF THE PROBLEM AREAS

As mentioned in Section 12.1, the yard has already decided to move partially towards a product-oriented company, especially for container ships. This means that the design activities will be less customer-driven, and that in certain respects this case is similar to the case described in Chapter 9 (or even to the case in Chapter 10). An opposite design choice would be to strive for total customer satisfaction by a systematic integration of the customer into major activities in the product life cycle. Another design option open to the shipyard is to invest more substantially in production planning and control. This scenario is discussed in much more detail in Chapter 21. It could lead, for example, to using centralised rough planning and control systems (production coordination system) in combination with decentralised detailed planning systems for shop floor production planning and control.

Yet another design choice would be to invest more in CAD and/or CAM. At this point the question of the optimum level of computer integration arises. Due to its complexity, the individual differences between companies and the relatively small market for IT vendors, and in view of the complex functionalities for suitable software products, many of the requirements as described above could not be satisfied in the past. These functionalities include:

- Consistent use of the different 3D-CAD-systems during the design of the entire ship, aimed at complete substitution of physical engine room models.
- Consistent use of CAM systems with complex links to CAD for NC path generation and simulation of (for example) profile bending, welding robots for the first, second and third phases of the shipbuilding process (cf. figure 11.5).
- Integrated computer-aided engineering (CAE) and CAD systems allowing stage-wise operations planning (for calculation purposes and manufacturing planning).
- New concepts for user-friendly and real-time data presentation (e.g. monitoring manufacturing and assembly progress and visualisation).
- Neutral databases and database management concepts for product, factory and process data.
- Use of communication technology for temporary internal and external stationary and non-stationary applications throughout the entire life of a product (ship).

Some of these alternatives are discussed in a more general context in Part E of this book.

REFERENCES

1. Brodda, J., Shipyard Modeling - An Approach to a Comprehensive Understanding of Functions and Activities; *Journal of Ship Production*, Vol. 7, No.2, May 1991, pp. 79-93
2. Hassan, N.; Thoben, K.-D., Towards a structured methodology for the ship predesign process; in J.B. Caldwell and G.Ward: *PRADS '92 (Practical Design of ships and mobile Units)* Conference Proceedings; Elsevier Applied Science, 1992
3. Kuhlmann, T.; Lischke, C.; Lehne, M.; Wollert, J., *"Informationsflußanalyse"*; BMFT-Projekt "PPS auf Werften"; Project Deliverable; Bremen 1989
4. Kuhlmann, T., *"Grobpanungssystem"*; BMFT-Projekt "PPS auf Werften"; Final Project Deliverable, Part III.1a; Bremen 1992
5. MUSYK Consortium (Eds.), *Requirements Definition of the Coordination Modules*; Deliverable D41; Esprit Project 6391; Bremen 1992
6. Speer, B.; Puleo, S., "Anforderungsprofil für die integrierte Planung und Steuerung von Konstruktion, Materialwirtschaft, Fertigung und Montage in der Werftindustrie."; *FDS-Bericht No.177*, Hamburg Germany, 1986
7. ROCOCO Consortium (Eds.), *D12; Workpackage 1: Generation of system Architecture*; Task 12: Modelling of present situation ESPRIT Project 2439: ROCOCO; 1990
8. VDI (Eds.), *VDI-Richtlinie: Methodik zum Entwickeln und Konstruieren technischer Systeme und Produkte*; 1993

Part C
WORKFLOW AND RESOURCES

13

Introduction to Part C

As illustrated in figure 13.1, which is similar to figure 5.1, manufacturing consists of three different processes.
Production is concerned with the physical transformation of products. Raw materials are transformed in production into finished goods. *Engineering* is responsible for designing the product and defining the production process. This represents the technological basis for production by defining what and how to manufacture. Requirement documents are transformed in engineering into documents with technical specifications. Both these processes are controlled by the *management* process. This results in work orders and defines when, where and how much to produce or engineer. The flow through the management process is control information.

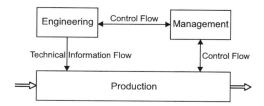

Fig. 13.1 Manufacturing processes

It may be observed that two of the processes (production and engineering) are operational whilst one (management) is managerial. The decomposition of the operational processes in engineering and production is necessary because they are executed at different frequency in the enterprise, depending (amongst others) on the customer-order decoupling point defined in the typology in Chapter 6. Downstream of the CODP both engineering and production are driven by customer order. Upstream of the CODP engineering is organized in a few larger projects whereas production is organized in a more repetitive way.

Workflows and resources are the central aspects of Part C. The workflows and resources are characterized by the layout of the physical resources, the categorization of the human resources and by the production and engineering processes.

The processes in Figure 13.1 cover any manufacturing typology. However, in this context the focus is on customer driven manufacturing. This type of manufacturing has some characteristics and problems that do not exist in the manufacturing of standard products. First of all, engineering is present for each customer order. Customer driven production must always start by defining or specifying the product. This introduces uncertainty in the operational processes. Price and delivery time must be quoted before the product and the process is detailed, defined or known.

In many cases engineering and production overlap, causing rework or change orders necessary as details become clearly defined. In customer driven manufacturing the production equipment needs to be more of a general purpose nature than in series production. And this also applies to human resources. Human competence must be more general to cope with different solutions than elsewhere.

As a consequence engineering is of special importance. The engineering process can be described by seven process steps (see figure 14.2). A key question is, how far in this process quotations should be carried. This question is concerned with both costs and time. The cost aspect should take into account that engineering work carried out for tenders may not lead to a successful contract. The time aspect is that the deadline for tender may be too short to provide full product and production specification. Bids are therefore built on a certain degree of uncertainty.

In addition to costs, the engineering process should be designed to meet requirements from production control. Two different groups of design principles may be applied:

- defining operations
- defining production phases and production units.

In design of the production, the layout of equipment is an essential task. Again the duality between workflows and resources arises since the two extreme types of layout are the job shop and the flow shop. Group technology is a tool that can help in this design.

Another important question in production is the balance between the use of machines and human resources. In customer driven manufacturing human resources cannot just as easily be replaced by machines as is the case in a highly automated flow line. The need for skill and improvisations is significant. The quality of the human resources may turn out to be a conclusive competitive factor for a one-of-a-kind manufacturing enterprise.

Human resources therefore need special attention in the design of a customer-driven manufacturing plant. The management of human resources is different in a workflow-oriented or resource-oriented system.

Management of human resources is of course different from management of physical resources. A system of machinery may be a "dead" or a "living-non-human" system. Such systems are passive or reactive, but never active as a "living human" system containing human beings.

The human resources are dependent on age unlike the non-human systems. Physical strength of a human is reduced by age, whilst experience is increased to give an example.

Design of an engineering and production system is dependent on a modelling tool. Actually models play a central role both with respect to resources and products.

Product models are well known in many types. They can represent the bill of material and the geometry of the product. In this respect the bill of material is of special interest. Most one of a kind product is derived from some base or generic variant. To cope with large numbers of product variants, generic bills of material may be useful.

In the preceeding the most important aspects of workflow and resources have been introduced. This includes engineering, production, product modelling and human resources.

Each of these aspects are discussed in more detail in the succeeding chapters.

14

Engineering in customer driven manufacturing

14.1 INTRODUCTION

It should be apparent from the analysis presented in Chapter 9 that the design of the customer driven engineering process is not optimal in the case, described there. There is no proper definition of the different engineering sub-processes and engineering documents in addition to the lack of a formal engineering approach. The customer driven engineering process is redesigned in this chapter. The basic objective here is to redesign the engineering process in such a way as to ensure that the new process will provide a good transformation of the customer's requirements into a product description which is feasible to produce.

The terms *designing* and *engineering* are often used interchangeably in the literature. In a number of publications, however, a clear distinction is made between these two terms. Engineering covers a much broader area in this case. Product engineering and detailed design are seen as processes which are part of the total engineering activity. The term *product description* will be used to refer to the output of the different customer driven engineering sub-processes.

The steps in the engineering process are covered in Section 14.2. The design requirements in connection with controlling the process are explained in Sections 14.3 and 14.4. The design of the engineering process is considered to be complete with the inclusion of these design requirements in Section 14.5, the final section in this chapter.

14.2 THE STRUCTURE OF THE ENGINEERING PROCESS

The engineering process is divided into phases of associated development activities in a structure model or a phase model for the engineering process. These activities each produce a product description at a certain stage of development. These successive product descriptions do not provide for alternative designs, but instead represent ongoing developments of the same product description which become increasingly detailed and concrete. There is a decision point following the completion of each phase; at this point the cumulative results are reviewed and evaluated before a decision is made whether to continue with the next phase. The basic development cycle is followed within each phase (see Figure 14.1) as the level of development becomes increasingly concrete. Phase models are based upon the concept that the product description being developed can be documented successively at three levels of abstraction, namely:

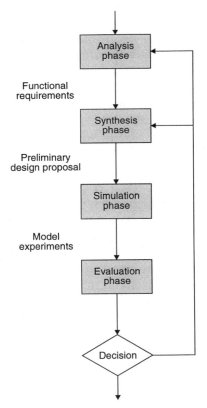

Fig. 14.1 The basic development cycle

- in the form of a functional structure;
- as a structured set of solution principles;
- as a fully specified technical product description.

An extremely large number of phase models for the engineering process or product development process based upon the above-mentioned levels have been proposed in the literature about product development. Examples of this have been published by Hansen (1976), Roth (1982), Pahl and Beitz (1988) and VDI 2221 (1986). These models all closely resemble each other and, for the most part, differ only with respect to the terminology used. The phase model proposed by the Verein Deutscher Ingenieure (VDI 2221 1986) is used here as the basis for designing the customer driven engineering process. This model, in particular, is cited frequently in the international professional literature.

The VDI phase model is presented in Figure 14.2, with terminology which has been modified somewhat to relate better to the customer driven engineering process. A total of seven concrete engineering steps and deliverables can be identified. The first engineering step is defining and analyzing the customer's problem. This results in a definition of the functional requirements for the custom-built product. This first engineering step is included in the problem analysis phase. The second engineering step involves defining the functions and sub-functions which must be included in the product. This results in a description of the functional structure of the product. The third engineering step concerns the development of solution principles. This results in a description of selected solution principles and alternatives which are related to the functional structure of the product. In the fourth engineering step, the sub-functions and associated solution principles are translated into feasible product modules which, together, form the modular structure.

The fifth engineering step provides for a general specification of the geometry and the materials to be used in the most important modules. This results in a preliminary draft of the product in the form of sketches and general specifications. The detailed geometry and materials are defined in the form of technical drawings and bills-of-materials during the sixth engineering step. The technical documentation for the equipment is also prepared. The manufacturing and assembly instructions for all of the product components and sub-assemblies are developed in the seventh and last engineering step.

162 *Engineering in customer driven manufacturing*

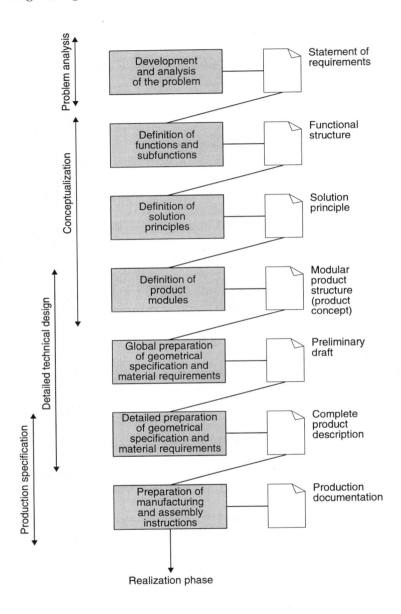

Fig. 14.2 Phase model for the engineering process (based upon VDI 2221 1986)

14.3 DESIGN REQUIREMENTS RESULTING FROM COST CONTROL

Requirements are associated with the design of the customer driven engineering process from a cost control point-of-view. During the customer driven engineering process, the first contact with potential customers concerns the possibility of supplying specific equipment (see Chapter 8). Since the equipment is customized and generally one-of-a-kind, an initial discussion is held concerning the equipment specifications and functions. Based upon this, the company provides a quotation to the customer. Since this type of equipment is normally seen as a capital investment, the potential customer typically requests quotations from various suppliers and then compares them. This means that there is a significant chance that the potential customer will end up ordering the equipment from a competitor. The average success rate for quotations becoming orders in this specific market is relatively low, approximately 15%. The financial consequences of this relatively low success rate is that resource capacity is utilized and costs are incurred in connection with a large number of quotations which generate no revenue. This can be seen as a certain financial risk. From a cost control point-of-view, an important design parameter with respect to the engineering process is the determination of which engineering steps are to be carried out during the quotation negotiation phase and which engineering steps are to be carried out after a firm order has been received.

14.3.1 Differentiating between the quotation phase and the order phase

Two extreme alternatives are conceivable in connection with separating the quotation phase from the order phase. One alternative is that only the problem analysis step, and thus no other engineering step, is performed during the quotation phase. The significant advantage of taking this approach is that only a limited amount of resource capacity is spent when there is a possibility of no revenue being received to cover this expense. The other engineering steps are then carried out only after the customer order is received. The major disadvantage of this approach, however, is that a quotation is made which commits the company to a price and a delivery date for the equipment, but based upon extremely limited knowledge of the product to be produced.

Committing to a price and a delivery date in this type of situation involves significant technical, time and financial risks (see also Chapters 9 and 12). With the other alternative, all of the engineering steps are performed during the quotation phase. The major advantage of this approach is that the complete product description is completed before the quotation is made. All of the product uncertainties are eliminated in this way. The technical risk and the associated financial and time risks involved in committing to a price and a delivery date are, thus, much less in this situation. This alternative has two major disadvantages, however. In the first place, a large amount of (scarce) resource capacity is utilized to perform all of the engineering steps during the quotation phase, with a significant risk that the quotation will not result in an order. In the second place, performing all of the engineering activities during the quotation phase would require a throughput time which would be unacceptably long for just providing a quotation.

14.3.2 Designing the quotation phase in the case situation

Both of the alternatives have major disadvantages in addition to their advantages. In view of the major disadvantages associated with both of these alternatives (a large risk versus the use of scarce resource capacity and a long throughput time), neither of these alternatives are acceptable. The design approach chosen for the case described in Chapter 9 needs to lie somewhere between these two alternatives.

Given the price level and delivery time restrictions, a sufficient number of engineering steps (but no more than this) need to be carried out during the quotation phase in order to be able to estimate the resource capacity requirements and costs with an acceptable level of risk. To estimate the material and manufacturing costs, the product description must be worked out to, minimally, the product concept level and the basic decisions regarding product geometry and composition of materials must have been made for the most important modules. In view of the chosen level of standardization in the case situation, it makes sense to perform the first five engineering steps (see Figure 14.2) during the quotation phase. This means that the product description is developed through to the preliminary draft in the quotation phase.

14.4 DESIGN REQUIREMENTS RESULTING FROM PRODUCTION CONTROL

14.4.1 Design principles for structuring a transformation process

Requirements for structuring the customer driven engineering process can also be formulated from a production control point-of-view. For structuring the process for production control purposes, design guidelines will be used which have been developed for the production control of *physical* transformation processes. We will demonstrate that these principles can be adapted easily and used effectively for structuring the production control aspects of customer driven engineering. A number of basic design principles for structuring a physical transformation process, borrowed from Bertrand et al. (1990), will be explained here.

We can distinguish two groups of design principles:

- defining the operations;
- defining production phases and production units.

Defining operations

A transformation process can be viewed as a set of production steps which must be completed in a certain sequence. Input materials and manufacturing resource capacity are both required to carry out each production step. The purpose of the production control is to coordinate all of the production steps included in the transformation process by assigning material and resource capacity to each of the production steps. It is sensible to aggregate a group of sequential production steps into a single "operation" when the same resource capacity is used and there are little or no intermediate waiting times. The production control problem can be simplified in this way since the total number of activities to be coordinated is reduced.

Defining production phases and production units

Even after the production steps have been aggregated into operations, the production control of the different operations is still typically a rather complex matter. The option of assigning all of the control activities to a single, centralized functional unit remains impractical.

The complexity of the control requirements can be reduced further by aggregating operations into production phases. Subsequently, the production phases are assigned to a so-called production unit (PU). A PU is an organizational grouping of resource capacities with the following characteristics (Bertrand and Wijngaard 1986):

- internally organized such that the operations, which are required to complete a given production phase can be performed independently, provided that the required materials and resource capacities are available;
- capable of making reliable commitments with respect to the specific conditions (such as utilization levels, throughput times, etc.) under which the operations belonging to a given production phase for a specified volume and for specified periods of time can be performed.

In this way the control problem is solved at two levels: the production unit control (PUC) level and the goods flow control (GFC) level. The use of GFC provides only for the coordination of the production phases in the transformation process. At this control level the total transformation process is defined only in terms of a set of production phases with relationships between these phases. Each production phase results in the delivery of a specified (intermediate) product which is referred to as a GFC item. Coordination of the production phases at the GFC level is accomplished by releasing work orders to the PUs to carry out specific production phases to produce GFC items. A work order is therefore an instruction to initiate and complete a specific production phase. Execution of the work orders is subsequently controlled at the PUC level where the performance of the various work order operations are coordinated within a PU. Each individual PU is responsible for completing all of its work orders in accordance with the agreed conditions. It should be clear that the GFC cannot release work orders to the PU arbitrarily.

The selection of production phases (and thus the GFC items) and the establishment of production units (by assigning production phases) are two explicit steps which need to be taken with respect to structuring the transformation process. A number of criteria can be identified for use in determining the GFC items and production phases as well as for establishing the production units (by assigning production phases).

Criteria for determining the GFC items and production phases

Reducing uncertainty The first criterion is the reduction of uncertainty. Whenever a significant reduction in uncertainty occurs at specific points in time during the transformation process, then it is advantageous to define GFC items at these points. In this way an opportunity can be created for reacting immediately to updated information at the GFC level. It is also desirable to create a buffer at each of these points so that the subsequent phase can be isolated from the uncertainties which exist in the previous phase. Examples of uncertainties in this sense are uncertainty of demand, uncertainty of yield, uncertainty with respect to resource capacity utilization and the uncertainty with respect to throughput time. By defining a GFC item directly following an operation for which the yield is uncertain, the actual yield information will be reported at the GFC level at the earliest possible point in time. In this way the updated information can be used to make adjustments as necessary to the production plan for the subsequent operations by changing the work orders and/or increasing safety stock or allocating additional production capacity.

Presence of a resource capacity bottleneck The second criterion is the presence of a resource capacity bottleneck in the transformation process. A capacity bottleneck can be seen as a limiting resource for the transformation process which determines the maximum quantity of output per unit of time for the process. For this reason, it is of utmost importance to ensure an optimal utilization of the resource capacity at this bottleneck by controlling the release of work orders prudently at the GFC level. To achieve this, a GFC item should be defined immediately prior to the capacity bottleneck. In this way the most up-to-date information on progress will be available at the GFC level to determine which work orders should be treated with a higher priority or if capacity should be utilized to increase intermediate stock levels. A GFC item should also be defined after the bottleneck in the event that not all of the work processed at the bottleneck will be needed immediately by the subsequent operation in the transformation process.

The product structure The third criterion is the product structure. Some components or sub-assemblies may be more critical than others for the overall coordination of the flow of goods. This is especially the case for situations in which complex assembly processes are required (for example, situations in which phantom parts are used in the MRP literature). When certain sub-assemblies are processed before others, this generally means that certain components will not be needed until a later point in time in the assembly process.

The three criteria described above are strongly oriented toward physical transformation processes and are, thus, not completely suitable in this form for defining GFC items within non-physical transformation processes. Nevertheless, specific criteria can be derived as special cases of the above-mentioned criteria; these criteria can then be used for defining the GFC items in the customer driven engineering process. In contrast with physical transformation processes, the problems of an uncertain yield and an uncertain product demand are not found here. The uncertainty which is foremost here with respect to the control aspects is the uncertainty about the exact product specifications, otherwise known as the product uncertainty. The product uncertainty decreases continually during the course of the engineering process, however, as the product specifications are defined in more detail. This information is updated and reported to support an integral control and correction of the total transformation process as soon as a significant part of the product uncertainty has been eliminated. The exact resource capacity requirements and the specific purpose of the various engineering steps are not known in advance. Information is only available in terms of which generic engineering steps need to be followed and which generic engineering documentation needs to be produced.

It is not just the resource capacity requirements which are difficult to determine during the engineering process. In addition, the resource capacity requirements for the physical production phase are not known and can, thus, only be estimated. Nevertheless, as the product uncertainty is reduced, a better estimate of the resource capacity requirements can be made for the remaining transformation processes (including the physical production). As better estimates are made, the previous control decisions can be evaluated and/or new control decisions made to redirect the remaining transformation processes as necessary. At these points in time, GFC items are needed in the customer driven engineering process.

The criterion of the presence of a resource capacity bottleneck is directly applicable to the customer driven engineering process and the definition of GFC items in connection with this.

The product structure in connection with the customer driven engineering process (or, in other words, the structure of the product documentation) is relatively simple when compared with a physical transformation process in production situations discussed here. There is no complex "assembly" activity. This means that the required GFC items are relatively easy to identify and define. Even so, this criterion is significant and applicable to customer driven engineering due to the important sequential relationship between the customer driven engineering process and the subsequent physical production. In the case from Chapter 9, in particular, a complex assembly process is included as part of the physical production of the manufactured equipment. A certain sequential relationship is inherent in carrying out the various assembly processes. Certain sub-assemblies must be completed in advance of others. These sequential relationships have an important effect on the priority of carrying out certain engineering activities. This can lead to a situation in which similar engineering activities for a given customer order are allocated to different engineering orders within the engineering process. This leads to different GFC items.

Criteria for establishing the production units
Production units are established based upon the resource capacities needed for the processing in the production phases. Multiple production phases may be organized in such a way that they are included within a single production unit. In the event that a separate production unit is created for each production phase, there will be an equal number of production units and production phases. This would generally be undesirable, however. In many situations it is likely that all of the resource capacities required by a number of different production phases will be included within a single production unit. In this case, all of the production phases assigned to a given production unit do not need to be linked together in the primary transformation process. We have identified two criteria for establishing production units by combining resource capacities (Bertrand et al. 1992b), namely:

- the nature of the required capability;
- the nature of the control issues.

Nature of the required capability The practical establishment of a production unit is dictated partly by the particular technical and functional aspects of the primary transformation process. So-called "production departments" are typically established as the result of functional specialization, economies-of-scale and technical similarities. A production department is therefore an organizational grouping of resource capacities which are closely related with respect to the nature of the available capability. It is sensible to define the PUs along the same lines as the production departments whenever possible. This does not mean that a production department should always be defined to be the same as a PU, however. The PUs are typically product-oriented rather than operation-oriented. PUs are not based upon the functional specializations within the existing production organization. Normally, it does not make sense to create a PU and to assign production phases to it which require different and dissimilar skills and capabilities. This could lead to a resource capacity situation within the PU which is not easy to model at the GFC level, making it impossible to coordinate the aggregated resource capacities. The nature of the capabilities represented within a PU should be related in some way. The use of this criterion in practice may cause similar activities for different product families (e.g., engineering activities) to be assigned to different PUs when the required capabilities for one product family are significantly different than the required capabilities for a different product family. A clear product orientation may develop in this way.

Nature of the control issues The complexity of the material coordination and the capacity coordination activities determines the complexity of the production control of a PU. The nature of the control issues may change at different points in time as the goods flow through the primary transformation process. In view of the different nature of the control issues, the control approach will be different in different production situations. Changes in the production situation can usually be identified as fixed points in the process. One example of this is the transition from manufacturing components to the assembly of components. In most situations involving the manufacturing of engineer-to-order equipment, the manufacturing of components can be characterized as a Job Shop situation while the assembly of components can be characterized as Project Assembly. It does not make sense to assign two production phases which require different control approaches to the same PU.

14.4.2 Applying production control design principles to the engineering process

We shall apply these design principles to the engineering process of the case situation described in Chapter 9.

Defining the operations

The selected operations are named in each engineering step appearing in Figure 14.2. An operation in the customer driven engineering process has a number of specific characteristics in comparison with an operation in a physical transformation process. In the first place, the input information in the customer driven engineering process is transformed to other information (the output) while the input information remains intact. In this sense, new information is created. This is different in the case of a physical transformation in which the input materials are physically transformed into a material with different characteristics. In the second place, activities which take place within a given engineering step generally make use of a single type of resource capacity, which is typically one specific person. These activities can be combined into a single operation if no discernable delay is encountered between them. In this way an operation in a customer driven engineering context may comprise many more different activities than an operation found within a physical transformation process. In the third place, a distinction can be made between (creative) information-generating activities and (repetitive) information processing activities. It is difficult to estimate the actual resource capacity requirements for the first category, in particular. We assume that every information-generating activity also implies the existence of one or more related information processing activities. In this way, the latter category can be viewed as an integral part of the first. In contrast with a physical transformation situation, it is possible that an engineer is busy with a single engineering activity for more than one machine at the same time. For the purpose of our model, an engineering operation is defined as a related set of engineering activities which are carried out sequentially. This assumes that an engineer first completes an engineering operation for one machine before he starts an engineering operation for a different machine.

The following operations can be identified:

- problem analysis. The most important operation associated with this step is preparing the specification of the customer's functional requirements;
- determining the functional structure and the sub-functions of the machine. This can be combined into a single operation of developing the functional structure;
- developing solution principles. This may require different capabilities, depending upon the types of sub-functions;
- grouping sub-functions and solution principles into product modules. Determining the modular structure of the product is partially supported by the availability of standard product modules.
- general specification of the geometry and materials for the most important modules; this can be seen as a preliminary draft of the product in the form of sketches and specifications;
- determining the details of the geometry and the materials to be used for all of the components of the machine and documenting them in technical drawings and bills-of-materials (detailed design);
- preparing the manufacturing and assembly instructions for the non-standard components and sub-assemblies.

During the sixth engineering step, the assembly structure of the machine is taken into account. The machine is split up into assembly modules. Sequential relationships are defined between the assembly modules which indicate the assembly priorities.

Defining the GFC items and production phases
Six different GFC items can be defined for the customer driven engineering process based upon the criteria described above. The first GFC item is defined as being the customer's *functional requirements specification* which documents the details of exactly what the customer wants. Based upon this, it should be determined whether it might be better to not comply with the customer's request for a quotation and, thus, not issue a quotation to the customer due to certain reasons (e.g., technical or business reasons) before time and effort is spent on the preparation of such a quotation (refer to Chapter 22 for a further explanation of this decision).

The second GFC item is the *preliminary draft* of the design of the machine, including the modular structure. A significant portion of the product uncertainty has been eliminated at this stage. A general estimate can be made with respect to the degree to which standard product modules can be used and, thus, the number of customized elements which will be included in the product description. The preliminary draft of the product is, in principle, the level of engineering required for preparing a quotation.

The third GFC item is a *generally detailed assembly module*. There are two reasons for defining a GFC item at this stage. In the first place, it is not sensible to view the complete detailed design of all of the assembly modules belonging to a given machine as being a single GFC item due to the sequential relationships between the assembly modules. These sequential relationships have an important effect on the priority of detailing certain assembly modules. The assembly modules with components which are defined as being on the critical path are detailed first. For this reason, it is useful to issue separate work orders for detailing the various assembly modules. In the second place, the general detailing of an assembly module provides insights into the resource capacity required for detailing the components of an assembly module. A significant portion of the product uncertainty is eliminated in this way.

This leads to the fourth GFC item, *the description of a set of detailed components* belonging to a single assembly module. The fifth GFC item is the *set of production documents* for the manufacturing and assembly of an assembly module.

The sixth and last GFC item is the preparation of the technical documentation for the machine. This activity is not included on the critical path for the engineering activities. As a result, it is normally completed just before the machine is delivered.

Creating production units

Five production units can be created based upon the production phases and criteria defined in Sub-section 14.4.1. The sales organization is responsible for soliciting requests for quotation from prospective customers. The sales organization also works out a functional requirements specification for each request for quotation which is received from a customer. This is used here as the starting point for the customer driven engineering process in the case. As a result, managing the sales organization and producing the functional requirements specification are not included as part of the customer driven engineering in this book and are, as such, not included as an aspect of the control issues discussed here.

The production phase "Complete the preliminary draft of the product" is performed within a single production unit. Since the nature of the required capabilities is quite specialized, this activity is, in principle, carried out by a single person. This production unit is called *Product Engineering*. The persons employed in this production unit are referred to as *product engineers*. A distinction is made between the product families, however. The required capabilities for primary packaging (bottling), on the one hand, and secondary and tertiary packaging, on the other hand, are radically different and cannot (or almost never) be shared. Therefore, two production units have been defined, namely: Product Engineering - Bottling and Product Engineering - Packaging.

The production phases of "Detailing the assembly module" and "Detailing the assembly module components" are carried out within a single, separate production unit since the nature of the required capabilities is the same for both phases. This production unit is called *Detail Design*. The persons employed within this production unit will be referred to as *detail designers*. Analogous to the situation described above, a distinction is also made here between the product families so that two production units are required, Detail Design - Bottling and Detail Design -Packaging.

The production phase "Compile the assembly module production documents" is carried out separately within a single production unit in view of the specialized nature of the required capabilities. There is a strong focus here on technical production aspects which is not found in the previous production phases. In addition, the control issues and the elements of uncertainty are minimal in comparison with the previous production phases. This production unit is called *Process Planning*. The persons employed within this production unit will be referred to as *process planners*. There is no requirement for making a distinction between the product families within this production phase in view of the fact that the process planners are multi-skilled.

By defining operations, GFC items, production phases and production units in this way, the production control design requirements have been observed in structuring the customer driven engineering process in the case.

REFERENCES

1. Muntslag, D.R., *Managing customer order driven engineering*, Ph.D. Thesis, Eindhoven University of Technology, 1993.

2. Bertrand, J.W.M., D.R. Muntslag and A.M. van de Wakker, De besturingsproblematiek in engineer-to-order bedrijven, *Bedrijfskunde* 64(3), pp.212-222, 1992a.
3. Bertrand, J.W.M., D.R. Muntslag and A.M. van de Wakker, De produktiebesturing in engineer-to-order bedrijven, *Bedrijfskunde* 64(3), pp.223-240, 1992b.
4. Bertrand, J.W.M., and J. Wijngaard, The structuring of Production Control Systems, *Journal of Operations and Production Management* (6), pp.5-20, 1986.
5. Bertrand, J.W.M., J.C. Wortmann and J. Wijngaard, *Production Control, A Structural and Design Oriented Approach*, Elsevier, 1990.
6. Hansen, F., *Konstruktionswissenschaft; Grundlagen und Methoden*, Verlag Technik, 1976.
7. Pahl, G. and W. Beitz, *Engineering design; a systematic approach*, Design Council, 1988.
8. Roth, K., *Konstruieren mit Konstruktionskatalogen*, Springer, 1982.
9. VDI 2221, *Methodik zum Entwickeln und Konstruieren technischer Systeme und Produkte*, Verein Deutscher Ingenieure, 1986.

15

Production in CDM

15.1 INTRODUCTION

In this chapter we will focus on the organisation of the shop floor in customer driven manufacturing. According to the theoretical framework presented in Chapter 5, the shop floor is described in terms of:

- the resources (machinery, human operators, halls, cranes, software)
- the workflow (both the physical workflow and the procedures determining the information flow)
- the organisational/decisional structure.

Below, in Section 15.2 we will discuss the resource structure. We will focus primarily on physical resources (layout and materials handling systems), and introduce concepts such as flow shops, job shops, dock production, and intermediate forms. We will also discuss human resources (just as an introduction, however, because Chapter 17 is dedicated to human resources).

Next, in Section 15.3 we will discuss the workflow structure. We will introduce a typical workflow-oriented technique, Production Flow Analysis, which is the basis for Group Technology. This technology illustrates the relationship between workflow and resources: it shows that design choices on the structure of the workflows should match the design choices on the organization of the resources.

The organizational/decisional structure of the shop floor will be mentioned only shortly, because shop floor management is discussed in detail in Chapter 23.

The discussion in this Chapter is to a large extent independent of the production technology. However, we present some considerations regarding the choice of technology in Section 15.4. This Chapter is concluded in Section 15.5 with the remark, that techniques such as

PFA are very valuable for non-physical production as well — even in a manufacturing company.

To illustrate the relations between the three subjects we describe here the so-called 'focused factory'. The focused factory can be considered as a prototype for customer driven production in small enterprises. As such, it has been an inspiring example for larger enterprises on questions such as how to decentralize, and how to obtain "intrapreneurial" organizations.

On the road towards the focused factory, the concept of a *production unit* may be helpful. This concept was introduced in the previous chapter, in the context of structuring the engineering work. It will play an important role in the remainder of this book. We will introduce production units in the context of physical production after the focused factory.

The Focused Factory

A 'focused factory' (Skinner, 1974) is a small factory with the following characteristics:

- There is a clear mission and everyone knows the essential aspects of sales, manufacturing, design and manufacturing technology. The communication within the factory is excellent.
- Management is located near the shop floor.
- There is little staff. Staff work near the shop floor, and staff employees are familiar with customers, shop floor workers and suppliers.
- There is a little or no specialisation among the management and staff. The managing director of the focused factory is also responsible for management of product design, process design, sales and financing.
- Shop floor workers carry out the standard maintenance of the machinery and are responsible for the quality. Specialised maintenance and quality engineers are under the supervision of the production manager, and often work in shifts.
- There is a strong involvement with the end product. Informal meetings are set up immediately in case of problems.
- There is little capital available for investments and the utilisation of the shop floor space is optimal. There is little on-hand stock, the lay-out is optimised, and if possible, one uses second hand machines with high technology.

This focused factory can be seen as the ideal model for a manufacturing organization. It stresses the autonomy of a factory with respect to resources, workflows, and decision making. It presupposes empowered employees. As such, it represents an important design choice for customer driven manufacturing.

However, the organisation of a medium or large enterprise according to this model is not that easy. The focused factory may be more a target than a reality, and a target which will not always satisfy all performance indicators. Thus, the requirements for production units are much weaker than for focused factories. Therefore, it is more easy to define and design production units in practice.

The production unit

A production unit is a (part of a) factory with:

- the ability to perform the operations of a given production phase independently
- the capability to make reliable delivey commitments (Bertrand and Wijngaard, 1986).

In other words, a production unit is defined as a group of people, production means and procedures that can transform autonomously specific raw materials to specific products, and it is possible to make agreements about quality, lead time and costs. Usually, there will be an inventory point (i.e. a goods flow control point) before and after a production unit.

The reason to introduce the concept of a production unit is that it enables us to decompose factories into smaller, autonomous parts. In real factories, the production units are often identical with departments. However, the choice of production units is one of the most important design choices, and it is often wrong to consider production units identical to the notion of a production department. A department is not a notion from the resources view and the workflow view, but from the organizational view. Preferably, organizational structures should folllow resources and workflow structures, rather than vice versa.

Although there is a clear distinction between the organisation of the shop floor and the management of the shop floor, they both influence each other. A specific lay-out can for example simplify the management drastically, as for example in the case of a lay-out for Just-in-Time production. On the other hand, admittedly, a certain type of control can determine a specific lay-out. JIT control requires for example a flow lay-out.

15.2 THE RESOURCES STRUCTURE

15.2.1 Job shop and flow shop

Functional versus product-oriented structure. A well-known distinction when setting up the shop floor is the distinction between a functional structure and a product-oriented structure. The term 'structure' can refer to the lay-out of the physical resources, but also to the organisation of the human resources. Some classical knowledge on the design of layout in general is given in the Appendix of this Chapter.

A *functional* structure is a structure where similar resources are brought together. This structure simplifies the management of the resources in terms of maintenance, utilization of the resources, standardisation, investment policy, etcetera. However, a functional structure leads to complex workflows. Therefore, internal logistics becomes difficult.

A *product-oriented* structure refers to a situation where resources are organized according to the workflow. This means essentially that different resources are put together in a way which enables them to produce a final product. The structure simplifies internal logistics, but it becomes more difficult to take advantage of similarity in resources.

In practice, one refers often to the distinction between job shop and flow shop. In this section we will discuss a number of issues that are related to the design of the organisational structure on the shop floor. It will appear that the most interesting forms of organisation in practice will be somewhere between the pure job shop and flow shop.

Job Shop. A job shop is a production unit organised as a number of work centres. The same type of machines within a work centre are set-up in parallel, and the workers in the work center are specialists on their machines. Products are transported criss-cross on the shop floor between the work centers, following a route that has been defined by work preparation (process design). The classsical functional organization is perhaps best described in the writings of F.W. Taylor on scientific management. In the classical functional organization, the blue-collar workers do not have indirect tasks. This means that task-elements such as quality control, costing, machine-maintenance, tool set-up and maintenance, work-prepararation, training, and many others are asssigned to specialist staff employees. In general one can state that shop floors which have a job shop structure, can produce a great variety of products. This is often called the flexibility of the shop. Within a limited scope, efficiency is also easily obtained in the job shop, because similar resources are combined. It is therefore unlikely that one resource is idle while there is a queue of work in front of a similar resource.

However, it is difficult to control logistics and quality in the job shop since the product flow is not standardised. As for logistics performance indicators, such as mean throughput time in the shop or due-date reliability, there are two reasons for this difficulty. First, there is no physical visibility of a "flow". It is not clear whether a workorder has just started or is almost finished. It is not clear whether a workorder is on time or not. It is not clear which measures have to be taken to keep logistics performance indicators under control. Therefore, shop floor planning systems have been designed mainly in job shops. Second, the natural orientation of the first-line supervisors in the job shop is concerned with resources — not with the workflow. It is natural that these supervisors make sure that their machines are in good shape and that their personnel are well trained. It is possible to control the efficiency and productivity of these resources by easy, objective measurements. It is much more difficult to exercise the same amount of control for logistics, because the logistics performance is the result of the joint effort of all first-line supervisors of a shop.

The case for quality control is similar. As we will see shortly, logistics and quality are highly intertwined. Here we may state that quality of the individual operations in a job shop is well guaranteed. However, this does not mean that the quality of the final product is under control. The focus in job shops is on the individual operations, not on the product as a whole.

This observation also has some consequence for the efficiency performance. Although the efficiency of each operation is well controlled in a job shop, the joint efficiency of the shop is often threatened. This is due to the fact, that the individual first-line supervisors and the workers tend to overemphasize their own quality. Often, procedures are followed, which are not necessary and therefore too expensive for a particular product. Furthermore, the job shop often shows super specialization. This means that although resources in a work center are similar, they are not the same, and they do not have joint queues of work. We will return to this point in Chapter 17.

Flow shop. A flow shop is a production unit organised as a number of of (flow) lines. Each line consists of a number of man/machine combinations called: *stations.* The basic principle of a line is, that all workorders have the same sequence of operations (routing), and that all operation times are the same. This leads to the concept of *cycle time*: the time which is available for subsequent operations on the same product by different stations, and also the time which is available for subsequent operations on different products by the same station. In contrast to the job shop, different machines rather than similar machines are grouped together in a flow shop, and are placed within one line in a certain order. Workers work within a line and, more specific, often on one station only. Products are manufactured according to the lay-out of the line and the order of the stations. There is therefore little flexibility in routings.

In flow shops, the workflow-orientation dominates over the resource-orientation. It is therefore easy to control logistics, but more difficult to optimize the utilzation of humans and machines. Furthermore, flow shops with very short cycle times are not suited for humans.

The classsical flow shops are Henry Ford's automotive factories. Charlie Chaplin's *Modern Times* is still an unsurpassable parody of the assembly line belt. It demonstrates clearly, that this production concept with its short cycle times leads to bad quality of working life.

Although logistics control is —in principle— very easy in flow shops, there are quite a few prerequisites. Similarity in routing and equal operation times have been mentioned above. However, excellent quality is another prerequisite. If the products have to be repaired or if a significant number of products fail, the efficiency and logistics performance are immediately damaged.

(N.B. The reverse is also true. In production systems with long leadtimes, quality control becomes more difficult because there is a long feedback cycle on quality and therefore no learning takes place).

Flow shop versus job shop. It is remarkable that the way of working in both classical job shops and classical flow shops share many disadvantages. Due to too much specialization, the tasks of workers are very limited. Therefore, the flexibility and efficiency in job shops, which is in principle available, is often difficult to mobilise. Also, quality and speed, which is in principle available in flow shops, gets lost because workers cannot be held responsible for products.

The cause for this state of affairs is functional specialization of tasks in far too much detail (Taylorism). Specialization has a horizontal dimension and a vertical dimension. Horizontal specialization means, that a worker can perform only a small part of the total number of operations required for his product. Vertical specialization means, that many indirect tasks are taken away from the worker on the shop floor. These indirect tasks include work preparation, tool control, machine maintenance, sequence control, quality control. Counter measures for these problems are known as work structuring and will be discussed in more detail in chapter 17. Quality of labour (job rotation, job enlargement and job enrichment) is the central theme of this work structuring.

15.2.2 Other forms of production units

There are many other ways to organise production units. Starting from the concept of a short-cyclic flowline, it is possible to implement the line in a way which allows workers to have some variation in their processing speed. Buffers can be introduced, the belt can be replaced by other means of transportation which are controlled individually by the worker, and cycle times can be enlarged. One way to obtain this, is to let the worker move with the product in a U-shaped small line. Another way is to introduce small batches of similar products.

Another measure is to split a long flowline into a number of smaller lines (minilines) with buffers in between. This can be a step towards the introduction of autonomous groups in the factory, where such a group is responsible for a miniline. Ultimately, this may lead to a focused (sub)factory. If the worker moves with the product in a miniline, this is somtimes called a *street*. A street offers many possibilities for work structuring by having the worker himself involved in material supply, tool supply, quality control, etc. Note, that it is also possible to keep the product at a fixed place and have the workers move along the products according to the same principles. This is a form of *dock production*, comparable to other forms.

A third way to improve short cyclic flowlines is by introducing parallel lines, which can produce the same type of products. The main disadvantage of parallel lines is the investment in production and transportation means. However, within the same product family, some specialization becomes possible. When parallelization is taken to its extreme, *dock production* emerges again. Here, the products remain at a particular place, and the worker(s) can build the product again in a fairly autonomous way. An ingenious form of dock production combined with flow production on the shop floor has been described in Chapter 12, the shipbuilding case.

All the above alternatives presume, that there is some product family, which has sufficient volume of production (and sufficient homgeneity to be suitable for flow production. In product- and workflow-oriented companies, this will often be the case. In the next chapter, we will focus on the development of such product families. However, in resource-oriented companies the natural point of departure is often the job shop.

In the job shop too, it is possible to reorganize the shop in a way which leads to defunctionalisation. A suitable technique is Production Flow Analysis, which leads to a structure called Group Technology. Group Technology splits up a job shop into a number of groups, which resemble the above mentioned streets: a group consists of all resources necessary to take responsibility for a well-define set of products in the shop. Group technology provides many possibilities for work structuring in a job shop, as we will see.

15.3 PRODUCTION FLOW ANALYSIS AND GROUP TECHNOLOGY

Work centers organised in a functional structure are often grouped according to specific manufacturing technology rather than according to the workflow of the product to be produced. This will result in increased product flow on the shop floor between different work centers. Burbidge (1989) proposes to study and improve the goods flow in a production unit with his method 'Production Flow Analysis' (PFA). The goal is to maintain only a limited number of main routings between production units (groups) and to make such a group a production unit responsible for a complete product or component. In order to obtain this goal, PFA takes a typical workflow view: all routings currently availble in the shop are sorted and grouped in order to get some recurrent pattern. Usually, a limited number of main routings occur in a job shop. Each of these can be selected as a candidate for groups.

Although some 80% of the routings can be classified normally with little effort in a group, there are a number of odd routings which require further analysis. Often, these routings fall in one group, but require one or two operations from another group. PFA has many techniques for treatment of such routings, which basically investigate whether the operation involved is really necessary. In many cases such a routing uses old technology, or the product can easily be redesigned to avoid the operations in question. Sometimes there is just an error in writing. When this is not the case, it is sometimes necessary to bring into a group of operations some additional equipment of quite a different nature in order to get a self-contained "group".

This can often be done with simple means and measures. For example, one can include a welding station in a sheet metal workshop. If this requires certain craftsmanship, then a worker has to move as well. However, in general it will involve job enlargement, which will in general influence the quality of work in a positive way.

Production Flow Analysis is a first step in order to find out whether the work in a production unit can be structured. It may lead to groups of routings which are similar in terms of their operations. This is the basis for Group Technology, which is discussed below.

Group Technology. Group technology means that the products are grouped into classes in order to reduce varieties or to make use of commonalities between the products in a class. In process planning for example, one can use group technology to reduce the number of routings and to use preference routings. A second step is to cluster the routings within a production unit into groups which have similarity in manufacturing. This can result in a splitting up of the production unit into a number of groups (cells). Each cell will then be responsible for a specified class of products. The physical layout of the shop floor should follow the definition of these cells (see figure 15.1).

The disadvantage of this splitting up will of course be the required investments in machines or operator training. The advantage is however a significant reduction of internal transport and a drastic reduction of planning and control (overheads). In many situations, the products within a cell will be more uniform which provides more efficient use of tools, transport material, floor space, packaging material, etc. Thus, one creates mini "focused factories" within a production unit. Once Group Technology has been implemented, it becomes more easy to introduce workstructuring programs, as will be discussed in Chapter 17.

186 Production in CDM

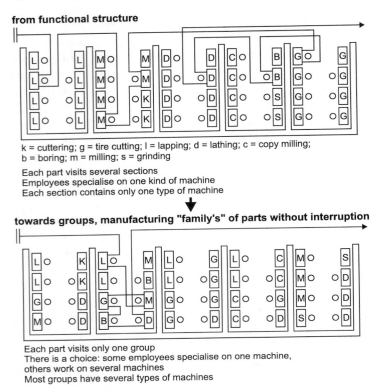

k = cuttering; g = tire cutting; l = lapping; d = lathing; c = copy milling;
b = boring; m = milling; s = grinding

Each part visits several sections
Employees specialise on one kind of machine
Each section contains only one type of machine

Each part visits only one group
There is a choice: some employees specialise on one machine,
others work on several machines
Most groups have several types of machines

Fig. 15.1 Group Technology

For the sake of completeness it is mentioned here, that PFA cannot only be applied at the level of the shop floor, but also at the level of factories and even of companies. At the level of companies PFA indicates whether the workflow between factories is well structured. At the level of a factories, PFA indicates whether the workflow between production units is well structured. At the level of a production unit, PFA indicates whether grouping in terms of cells is well structured.

15.4 DESIGN CONSIDERATIONS WITH RESPECT TO MANUFACTURING TECHNOLOGY

When setting up a factory and in particular when choosing the manufacturing technology to be used, one has to consider the distribution of tasks between man and machines. Many performance indicators play a role in the decision what technology to use. Some of these are:

- costs and expenditures
- short and reliable throughput times
- flexibility
- quality of work
- product quality.

Below we will discuss these aspects in some more detail. It should be noted here that the volume of production for which the production system is designed plays an important role in the choice of manufacturing technology. Especially in the area of customer driven manufacturing, these volumes will sometimes be low. This requires often universal machines with high flexibility, and it will certainly affect the trade-off between machines and skilled-human labour.

Costs versus expenditures. One can make a global calculation of the contribution to the total cost price when one knows the product cycle time and the life time of the machine. Although we will not present the economic techniques here, we can say that every technology will have its consequences for expenditures or savings somewhere else. For example, a technology that places high requirements on the raw material will likely increase the purchasing costs. Knowing that in many companies purchasing is responsible for up to 70% of the costs, it can be worthwhile to consider an alternative technology that imposes less requirements on the raw material.

Short and reliable throughput times. Short and reliable throughput times are mainly obtained by implementing some form of flow production. This includes that all process steps in the production process should have approximately the same cycle time, which reduces the choice of technology highly.

Flexibility. Flexibility involves many aspects of shop floor organisation. One can make a distinction between:

- volume flexibility: the extent to which and the time required to vary the production volume;

- mix flexibility: the extent to which one can vary the mixture of products within a product family, given a certain volume;
- product flexibility: the extent to which one can introduce new products on the existing production facilities.

Quality of work. With respect to the quality of work, it is important to design a production system that does not consider the human as an extension of the machine. The capabilities of the machines and humans have to be optimised together. This approach is indicated with the term sociotechnical approach. In the sociotechnical approach it is avoided to automate everything that technologically can be automated. Instead one tries to design tasks for man and machine that correspond to their capacities and capabilities which should be complementary. This will often result in cheaper solutions with an increased quality of work and more flexibility.

Product quality. Quality control and production control go hand in hand. Short throughput times result in shorter feedback from customers to suppliers. Also, a requirement for shorter throughput times and lower stock levels is that one can rely on the quality and that there will be no scrap.

15.5 CONCLUDING REMARKS

The considerations given in this chapter have great impact in the remainder of this book. The notions of focused factory, job shop, flow shop, and intermediate forms of production units will be used in Part D where production planning and control concepts will be discussed for the cases presented in Part B. The same is true for Group Technology. These production environments will also determine the IT-solutions to be presented in Part E, as well as the models which relate design choices to performance indicators in Part F.

However, these examples of production environments used in customer driven manufacturing are not the only message of the present chapter. Equally important is the insight, that both the classical flow shop and the classical job shop lead to narrowly defined tasks for production employees. This lack of quality of work is counterproductive. In Chapter 17 we will discuss in more depth what can be improved in the design of work systems.

Finally, it should be noted that the techniques and insights of this chapter can be applied easily to other work systems than only to the production floor. In fact, techniques such as PFA may be applied just as well to information processing organizations or service organizations, such as banks and hospitals. The term *Business Process (Re)Engineering* refers to this usage of techniques presented here. This observation has, in turn, important consequences for manufacturing organizations. More and more added value, time, and human activity is concentrated in non-physical workflows in manufacturing companies. Such workflows involve, for example, 'the customer-driven engineering process', discussed in the previous chapter. But also 'the way in which orders, tools, materials and documents are made available to the shop floor'. These business processes can be analysed and re-engineered in exactly the same way. Part E (especially Chapter 29) will give several examples of such re-engineering of non-physical workflows in customer driven manufacturing.

REFERENCES

1. Bertrand, J.W.M. and J. Wijngaard, The Structuring of Production Control Systems, *Journal of Operations and Production Management* (6), pp.5-20, 1986.
2. Burbidge, J.L., *Production Flow Analysis*, Oxford Science Publications, Oxford UK, 1989.
3. Richard Muther, *Systematic Layout Planning*, 1979.
4. Skinner,W., "The focused factory", *Harvard Business Review*, May/June 1974, pp.113-121.

APPENDIX TO CHAPTER 15: FACTORY LAYOUT DESIGN

APPENDIX 15.1 SLP AND SHA

The classic book in the area of factory lay-out is "Systematic Layout Planning" of Richard Muther (1979), which describes the method SLP. Muther also developed the Systematic Handling Analysis method (SHA) for the analysis of problems in material handling. SLP and SHA are based on the so-called PQRST data:

P = Product: what goods are handled?
Q = Quantity: how much is handled?
R = Route: whereto are the goods transported?
S = Service: by what means will the goods be transported?
T = Time: when will the goods be transported?

Although SLP and SHA are originally developed for the lay-out of warehouses, they are also valid for the lay-out of the shop floor.
There are however some new insights in the area of shop floor lay-out. Below we will summarise these concepts under the following headings:

packaging
line lay-out
internal transport
on-hand inventory

Packaging
Packaging of goods during transport (both internal and external) and in depots and warehouses is of great importance since it often determines: batch size, shop floor space needed, costs of transport, and working conditions and efficiency of workers on the shop floor.

Line layout
Also in customer-driven industry it may be an option to use the principles of flow production. The classical layout for flow production is a line layout. The term "line" refers to the physical layout of the production system. The basic shape of the line is the U-shape (see figure 15.2). Based on the U-shape one can create serpentines, which

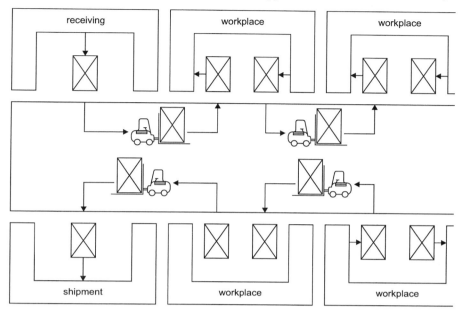

Fig. 15.2 Example of a U-shaped lay-out

can again create a U-pattern. There are four reasons to opt for a U-shape:

- to avoid return flows
- to increase the quality of work and the quality of products
- to stimulate team work
- to provide a flexible layout.

These arguments are elaborated on beneath.

First, a straight line has the disadvantage that for example empty pallets have to go back to the start of the line. This return flow requires extra space and transport which is avoided with a U-shaped layout.

Second, workers in the production line are less "static", and sometimes the worker has to follow the product on the line, and a certain range of operations is carried out by the worker on the product. This has great advantages in terms of product quality and quality of work.

Third, the U-shape reduces the distance between workers. This stimulates the team work between these workers on the same product. For example, inspection usually takes place at the end of the line. Efficient communication is of great importance in case of failures. This is more easy in a U-shaped layout.

Fourth, U-shapes result in a layout of the factory floor in more or less equal size squares. This allows more flexibility when changes have to be made.

The advantage of the straight line on the other hand is that the internal transport is more easy since there are no bends. Furthermore, reachability of the workers on the line is sometimes mentioned as a disavantage of a U-shaped layout, which is the case in a fully automated line that cannot be crossed.

16

Product Modelling and other functions

16.1 INTRODUCTION

One of the cornerstones of a Computer Integrated Manufacturing (CIM) environment is the product model. This product model is used to represent product information for all sectors of engineering as well as for all aspects of the life cycle. Geometric models (e.g. IGES, STEP) and bills of material can be considered to be product models. Application programs can make use of the formal structures of these models, e.g. for analyzing the strength of a prototype or for identifying the critical components in a bill-of-material. Other application programs support humans in understanding the product development process. These applications make use of the product model to gain a common understanding among the engineers of the products that need to be developed.

However, most of today's product models represent only a small part of the product information and are not suitable for representing product families, i.e., products which often have a large number of variants (cars, medical equipment, televisions, etc.). Especially in product-oriented companies which are customer driven, these factors play an important role (see Chapter 6).

This chapter focuses on product modelling techniques for representing such families. Particular attention will be paid to so-called generic product modelling; generic bills of material are a prominent example of this. The reason for focusing on the generic bill-of-material approach is due to the fact that this approach provides a solution for the problems of the Medicom case as described in Chapter 10.

Firstly, we will provide a brief overview and explanation of what is meant by product modelling in the context here. We show that most product models are devoted to capturing detailed information and various data elements associated with product development. It becomes clear that only a limited number of product modelling approaches include the modelling of manufacturing operations, logistic operations and/or services in their description of the product life cycle. Then, in Section 16.3, we identify some of the shortcomings of these product models by using the Medicom case described in Chapter 10 where there was a modelling problem concerning a proliferation of product variants. The generic product modelling approach is then introduced in Section 16.4, based upon the use of generic bills of material. It is shown how this approach allows for a transparent and non-redundant description of a product family, thereby resolving the specification problems described in Chapter 10. Finally, modelling the operational process using generic bills of material is discussed in Section 16.5.

16.2 PRODUCT MODELS: AN OVERVIEW

In manufacturing situations, the product model is typically used to define the various data elements which appear throughout the product life cycle. In this chapter we focus on the data which describe the physical structure of a product family. This type of data is used by computer-aided design (CAD) applications for designing and testing a product configuration of a product. Bill-of-material systems also use this type of data which is needed as the basis for manufacturing the products. Furthermore, this type of data can be used by personnel to gain a common understanding of the general construction of a product, thus, simplifying communications between the different functions in the product life cycle.

The following list provides an (incomplete) overview of application areas in which product models may be used. Some references to the literature are mentioned for further reading.

Product models may be used for:

- representing product geometry and features. This type of product model is used by CAD systems to define the physical shape of a prototype.
- representing assembly structures. This type of product model builds upon the previous type of model by describing how monolithic parts

are put together. Tolerance analysis, interface management and design visualisation are important features of this type of model. For further reading refer to [Nevins and Whitney, 1989] and [Mäntylä], 1990].
- representing functions. In the initial design phases, the function of a product can be decomposed into subfunctions. These subfunctions can be tested using special applications and then specified in terms of physical components [Kota and Lee, 1990]
- simulating behaviour. Product models for specialised applications such as FMEA, heat transfer, real-time robotics, etc. can provide early feedback on the quality of the design [Nevins and Whitney, 1989].
- supporting production control functions. A manufacturing bill-of-material can also be regarded as a product model [Erens et al., 1992]. Unlike an assembly model, parts are put together based upon the steps of a production process.
- exchanging information. Exchanging product data requires a neutral format in which the geometry and other aspects of a product are described. The STandard for the Exchange of Product model data (STEP) includes the elements of product data models which define the form of product data as well as the product data technologies such as descriptive languages [Express, STEP Part 11], conformance testing methods and implementation forms.
- reusing product information. Salzberg and Watkins [1990] found that design hierarchies (representations of the structure and function of the firm's products and processes) could be effectively reused for new product designs.
- concurrent engineering. Lindeman and his colleagues [Lindeman et al., 1992] argue that the concurrent engineering methodology relies upon the ability to share up-to-date engineering product structure data.

16.3 SHORTCOMINGS OF THESE PRODUCT MODELS

The aforementioned product models are used for the description of single products and often include a relatively large amount of detail. This approach to the representation of product data is satisfactory only when single product designs are to be developed. An example of this is the floor pan of a car: a part which is manufactured in bulk, but designed as a single product. Although the developed products are unique, some support is given in the product models for managing versions and reusing information for new products. However, no explicit support is provided for managing many variants of the product in this case (see Figure 16.1).

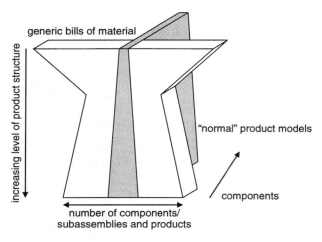

Fig. 16.1 Product variety versus product detail

As we have seen in Chapters 9 and 10, products may be available in millions of different configurations in customer driven and product-oriented companies. Medical equipment, for example, is assembled to customer order. This allows the customer to choose among hundreds of different options, resulting in millions of product variants. Similar examples can be found in the manufacture of cars, lorries, aircraft and lighting products.

Despite the fact that some products may have millions of variants, many of the variants will typically have a certain commonality. The term "family of variants" is used to denote the similarity of the product structure of these variants. This characteristic is exploited to model these variants using a single product model which does not have any data redundancy. Research into product variants has been carried out with respect to CAD systems as well as in the area of production control.

With respect to CAD, the concept of product variety can be found in the form of parametrised design [Yoshikawa, 1989 & 1990]. Parametrised design is not only used for modelling product families, however, it is also used to optimise the design of a given product. In both cases, parametrised design exploits anticipated geometric similarities in the designs. The differences are then parametrised in such a way that unique designs can be generated by populating the parameters with values. There are, thus, two basic types of parametrised designing processes:

- defining parameters for an artifact in order to find an optimal design. An optimal design can be found by varying parameter values, checking constraints on parameter values and simulating the functions of the prototype. Intermediate designs can be regarded as design versions since they are all based on common geometric characteristics. This type of designing process is also known as "constraint modelling" [Karandikar et al., 1992];
- defining parameters for an artifact in order to develop several variants in a transparent and non-redundant way. Each of these variants may then be viewed as a separate product. Products can also be grouped into product families to reduce the total design effort and to simplify manufacturing, servicing and other company functions. When parametrised designing is used to model product variants with respect to a given product assembly, the variants will share a common architecture while incorporating different components. An example of this could be greenhouses which are sold to customer order, but retain a common architecture while incorporating aluminium profiles of varying lengths.

With respect to production control, the problem of managing many similar bills of material often becomes cumbersome. The bills-of-material may be similar in the respect that a high percentage of the components are identical and, particularly, in the respect that they have a common structure. This is similar to the aforementioned architecture of parametrised design. The use of generic bills of material can be used to simplify the modelling of many similar bills of material in this type of situation.

16.4 MODELLING A RANGE OF PRODUCT VARIANTS

The Medicom case presented in Chapter 10 leads us to the conclusion that a solution for the sales, servicing, production control and manufacturing problems will require the development of two different product models:

- a "customer product model" for Commercial Operations which is also intended for use as the basis for all communications between sales and manufacturing. This model should be more stable than particular component specifications or finished product specifications.
- a "technical product model" for manufacturing purposes for System and Component Operations. This model should describe the finished products, subassemblies, components and their relationships.

The generic bill-of-material (GBOM) approach, mentioned in the previous section, can be used to combine the customer product model and the technical product model into a single, integrated product model. The objectives of this integrated product model are:

- to model a product family from both a customer's and a technical point of view;
- to model a product family (with its variants) without data redundancy;
- to support the use of applications for production control, e.g., based upon MRP-II concepts; and
- to improve product transparency to facilitate functions such as product management and development.

The use of a GBOM approach to achieve these objectives is based upon the following premises:

1. the customer's product specifications can best be described in terms of the functional characteristics of each product family and its variants;
2. the technical product specifications can be modelled based upon a definition of a product structure which is similar, though not identical, for all product variants;
3. the product variants within a product family are found at the lower levels of the product structure;
4. the customer and manufacturer will generally specify a product differently, but these different specifications can be mapped to each other within a shared product model;
5. a specific bill-of-material (product variant) for a specific customer order can be generated from the product family description.

These premises are explained in more detail in the following sections.

16.4.1 Functional characteristics and the customer's point of view

The customer's point of view described in this section is based on the functional characteristics of a product family and its variants. These functional product characteristics are:

- seen as an abstract of the technical component specifications;
- understood normally by sales & servicing personnel as well as manufacturing personnel; and
- more stable than the ever-changing component specifications and finished product specifications.

Functional product characteristics can best be defined using parameters and parameter values. Together, all parameters and parameter values should describe a complete product family from a customer's point of view. Constraints on these parameter values prohibit technically impossible or commercially unattractive product variants. The following abstract from a page of the new Medicom global sales catalogue is an example of such a functional product description for a "cardiovascular" product family:

```
System        Fixed         Plane           Voltage      Frequency
- cardio      - yes         - mono-plane    - 220V       - 50 Hz
- vascular    - no          - bi-plane      - 230V       - 60 Hz
                                            - 240V

Power         X-ray tube    # Monitors      Software
- medium      - WTZ 05/10   - 1             - standard
- high        - RDT 05/08   - 2             - extended
              - GFU 05/08   - 3
                            - 4

NOT ((Power=medium) and (Plane=bi-plane))
NOT ((System=vascular) and (X-ray tube=GFU 05/08))
```

Fig. 16.2 Sales catalogue

Such a catalogue page, also referred to as a "choice sheet", describes the scope of the cardiovascular product family in a clear, well-defined manner. In this sense, it provides a good basis for defining the "boundaries" of the product family, for example, when questions arise regarding whether a particular product variant belongs to a given product family. In the two other product specification approaches

200 *Product Modelling and other functions*

presented in Chapter 10, namely, specifying components and specifying preferred systems, much repeated effort was needed to determine whether a product variant could be designed and manufactured with a predictable cost and effort. In this case, the creation of a product family choice sheet helped to:

- resolve the issues concerning product family boundaries, prior to initiation of the actual manufacturing activities. The initial discussions which took place (as part of the change to this new way of specifying products) cost more than a half year because the existing product families were seen by Commercial Operations to be much broader than the interpretation of System Operations. These issues have now become part of the normal discussions in specifying the customer's requirements during the product design phase;
- stimulate discussions about defining product families for which all of the necessary component interfaces are considered beforehand. Defining product families proactively (instead of reactively based upon customer orders) requires, however, that the architecture of the product family is determined in an early phase of the design process.

Choice sheets are used in the operational process to specify a customer order specific product variant. This is accomplished by simply noting the relevant parameter values (see Figure 16.3). A computerised version of the sales catalogue will make it possible to automatically validate the entered data to ensure that the values fall within the predefined constraints.

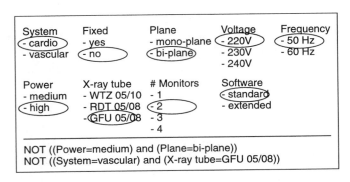

Fig. 16.3 Customer order specific product variant

16.4.2 Similar product structures and the technical point of view

The different variants within a product family have a similar product structure and are often produced on the same assembly line by the same workers and production machinery. This implies that the assembly and manufacturing operations have a repetitive nature, increasing the efficiency and productivity of the manufacturing process. At the same time, a range of product variants can be maintained to comply with customer and market demand.

A cardiovascular system, for example, is a product family with millions of possible variants. Nevertheless, all of these variants resemble each other to a large extent since they all include a stand, a table for the patient to rest on, an X-ray tube, a generator, an image intensifier and control software. The exact specifications for these components may, of course, be different for each of these product variants. The GBOM for this product family is defined in such a way that it accommodates all of the detailed differences in the variants. A simplified example is shown in Figure 16.4. Note that this product family structure corresponds to the top half of the hourglass which is not covered by the other two product specification methods (see Chapter 10). The component specification method describes only the lower part of the product family structure, while the preferred system specification method describes only a limited number of variants within the product family structure illustrated in Figure 16.4.

16.4.3 Product variants are found at the lower levels

A final product (such as a cardiovascular system) as well as the subassemblies and components of this product can be defined as product families provided that they all have a number of variants. A stand, for example, could be available as two different options: a cardiological version and a vascular version. Similarly, the X-ray tube might have different variants for different types of medical examina-

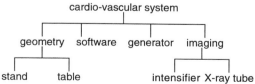

Fig. 16.4 GBOM for a medical system

tions. The definition of options for variants at the lower levels in the product structure provides the basis for a large number of product variants at higher levels in the product structure.

In terms of our hourglass model, the basis for defining product variants can be found in the neck of the hourglass; the resulting range of finished product variants is then found at the top of the hourglass. The component variants in the hourglass neck have specific bills-of-material, similar to Medicom's component specification approach as discussed in Chapter 10.

In an ideal situation, product variants are introduced relatively late in the manufacturing process. The shape of the product hourglass then becomes more of a mushroom since the product remains largely unchanged during the manufacturing process. The increased use of embedded software in the manufacturing process creates new opportunities for defining the functionality of a product by downloading software at the very last minute, making this mushroom even more pronounced.

16.4.4 Different customer and manufacturer views

When a customer orders a stand for cardiological examinations, he is normally not interested in a detailed technical specification of this requirement. In other words, the customer orders a "product feature" rather than the collection of physical parts which are used to implement that feature. The customer view is usually represented by parameters and parameter values as presented in the beginning of this section (see Figures 16.2 and 16.3).

On the other hand, the manufacturing view is based upon the physical structure of the product family (see Figure 16.4). As both the customer and the manufacturer typically have different ways of viewing a product family, a consistent mapping of one view to the other must be carefully maintained. A change in one view will normally be accompanied by a change in the other view. In order to create a generic bill-of-material it is necessary to determine exactly which parameters influence which (component) families. This is illustrated in Table 16.1 for the cardiovascular example.

Parameter	Component Families
system	stand, table
fixed	table
voltage	generator
hertz	generator, intensifier
power	generator
X-ray tube	X-Ray tube, intensifier
monitors	stand
software	stand, X-ray tube, generator

Table 6.1
Cardiovascular parameters and the component families they affect

The *system* parameter determines which of two generic component products are to be used as shown in Table 16.1. This type of parameter is best controlled at the first common parent of the component products which are involved. This means that the *system* parameter will be related to the geometry subassembly shown in Figure 16.4. On the other hand, the *voltage* parameter is related directly to the generator since this is the only component family involved.

The values of parameters such as the *system* parameter can be "inherited" and passed on through the product structure to the levels of the relevant component families. As a result, a distinction can be made between internal and external parameters for a given product. An internal parameter is a parameter that is defined locally for a specific product family, while an external parameter is defined at a higher product family level and inherited by the relevant component families. Figure 16.5 shows a subset of the physical structure and the parameters that are related to the families of this physical structure. For example, *system* is an internal parameter for the geometry family since it is defined at this level. For the stand and the table components, however,

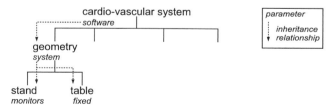

Fig. 16.5 Inheriting parameters

it is an external parameter since it is inherited from the higher level geometry family. Another example is the *monitors* parameter which is a local parameter for the stand and is not used by other families in the structure. Such a parameter is normally easier to control from both an engineering and a production control point of view. The *software* parameter, on the other hand, is defined at a higher level in the structure where it affects a large number of the medical system components.

16.4.5 Generating specific bills-of-material

A customer order specified in terms of parameters and parameter values can be interpreted by a generic bill-of-material system to generate a customer order specific bill-of-material. A generic bill-of-material incorporates a product structure which is the technical translation of the choice sheet used by sales. Each sales variant can be mapped to a technical variant by the generic bill-of-material system.

We have already explained how the range of product variants at higher levels in the product structure are derived from the product variants at lower levels in the product structure. The product families which are not broken down into further components in the GBOM and can be seen as the basic building blocks for the product variants are called generic primary products (GPPs). Their variants are called primary variants. All other families are called generic secondary products (GSPs) and their product variants are referred to as secondary variants. The stand, table, software, generator, X-ray tube and image intensifier are all generic primary products. The geometry, imaging subsystem and the cardiovascular system are generic secondary products.

The parameters associated with the GPPs are used to select the relevant primary variants for a customer order (in terms of parameters and parameter values). The GPP table, for example, has a *system* parameter and a *fixed* parameter.

The generation process for a secondary variant can be summarised as follows:
1. determine the value for each internal parameter related to a product family;
2. determine the inherited value for each external parameter related to the component families for which there are matching parameters at the parent level;

3. in the case of a GSP, create a customer order specific secondary variant or select an existing variant if it happens to be in stock;
4. in the case of a GPP, select a primary variant.

This process can be found in (commercially available) generic bill-of-material systems [Bottema and van der Tang, 1992].

16.5 SALES-MANUFACTURING COMMUNICATIONS

Using the approach described in the previous section, we can now specify products differently to improve the ordering process as depicted in Chapter 10. For the sake of clarity, we will first provide a summary of the case situation and review the problems which Medicom were facing before implementing this new approach. Further details about the communications between sales and manufacturing in assemble-to-order companies can be found in Chapter 20: production planning & control in AtO.

Sales: The first – and perhaps most important – step is to gain an accurate understanding of the customer requirements. The salesperson is expected to be an expert in the customer's application area as well as the technical aspects of the medical equipment manufactured by Medicom. In this phase, prices and delivery times are discussed only in general terms. However, once the exact customer requirements become clear – a process which may take up to a year to complete – representatives from the sales logistics need to be involved.

Sales logistics: This department is responsible for producing a clear description and specification of the appropriate product variant, including quantities, delivery times and prices for quotations as well as the final orders. The sales catalogue is used as the basis for translating the customer requirements into parameters and parameter values. Since there is agreement between Sales and System Operations with respect to the product information as it is structured in the sales catalogue, a minimal amount of additional communication is needed to ascertain whether a specific product variant can be manufactured. Furthermore, the price of the product can be determined easily by adding up the prices associated with each of the parameter values. In this case, the immediate availability of product prices was a major improvement over the previous situation in which prices were negotiated internally between Sales and System Operations for every single customer order. Delivery dates are still established only after consulting the responsible Product Group, however, a provisional delivery date can normally be assigned immediately based upon the information which is available at the family level. If product requirements lead to products which are on stock in a local warehouse, an alternative delivery date can be given to the customer.

Communicating order information: The information regarding new orders is communicated via the normal company infrastructure. Although no people are involved in this communications process, errors can still occur during the manual entry of data and in conjunction with printing the customer orders. The number of errors has been reduced considerably in the new situation, however, since the need for specifying hundreds of component codes each time has been eliminated. In addition, the range of parameter values has been reduced and, more importantly, these parameters are less sensitive to technical changes. The separation of product specifications into a customer's point of view and a technical point of view has resolved the problem of inconsistent product coding.

Order acceptance: The main purpose of the order acceptance function is to screen orders and to translate each order into a specific bill-of-material. This initial manufacturing function used to be a major bottleneck with respect to communicating the order information since each order based upon component specifications had to be checked thoroughly to determine if the specified combination of components was feasible from a manufacturing point of view. Lengthy and numerous communications with country sales offices were typically necessary to arrive at a product specification with the latest versions of the ordered components. Only then could the delivery lead time and intercompany invoice price be determined. The Commercial Operations personnel became frustrated while awaiting answers. In the new situation, however, order screening, order translating and delivery lead time confirmation can be accomplished within a day.

Final assembly: These manufacturing operations have benefitted considerably from the generic bill-of-material approach since the structure of a finished product variant is now known. In the original situation, when only bills-of-materials for primary variants were defined, the final assembly of product variants could not be controlled sufficiently. This introduced a certain degree of uncertainty about how primary variants were to be assembled into secondary variants (subsystems and finished products). An additional advantage of the generic bill-of-material approach is the enhanced possibility of moving the customer order decoupling point upstream, thereby transferring planning driven manufacturing to customer driven assembly. This typically results in a considerable reduction of the component and subassembly stock requirements.

Most of the aforementioned benefits are dependent upon an improved specification of the product families and product variants. This, in turn, improves the communication between the various links in the order chain. In addition, this improved product specification is shared and understood by sales as well as the technical manufacturing functions. This makes it easier to agree on the product family boundaries, prices and delivery lead times prior to the arrival of customer orders. As a result, the throughput time required for the operational processes is also reduced.

16.6 SUMMARY AND CONCLUSIONS

In this chapter, we have provided an overview of currently used product models and were able to conclude that these models are not suitable for use in a situation in which a large number of product variants exist. In Chapter 10 we argued that the key to improving the coordination of sales and manufacturing activities in the Medicom case was to develop a transparent, consistent and shared view of product information and specifications. This chapter introduced a new product modelling approach, called generic bills-of-material, which provides for separate views of the products for the purpose of communicating with the customer and for technical manufacturing purposes. This approach combines both of these views within a single framework. The case study shows that this approach can improve the communication between sales and manufacturing groups considerably, thereby enhancing the quality of information, logistic performance and customer satisfaction.

REFERENCES

1. Bottema, A. and L. van der Tang, A Product Configurator as Key Decision Support System, in: *Integration in Production Management*

Systems, H.J. Pels and J.C. Wortmann (eds.), Elsevier Science Publishers, IFIP, 1992
2. Erens, F.J., H.M.H. Hegge, E.A. van Veen and J.C. Wortmann, Generative Bills-of-Material: an Overview, in: *Integration in Production Management Systems*, H.J. Pels and J.C. Wortmann (eds), Elsevier Science Publishers, IFIP, 1992
3. Hegge, H.M.H. and J.C. Wortmann, Generic Bill-of-Material: a New Product Model, *International Journal of Production Economics* 23, pp117-128, 1991
4. Hegge, H.M.H., A Generic Bill-of-Material Using Indirect Identifications of Products, *Production Planning and Control*, vol. 3 no. 3, pp.336-342, 1992
5. Hegge, H.M.H., *Intelligent Product Family Descriptions for Business Applictions*, Ph.D. Thesis, Eindhoven University of Technology, 1995
6. Karandikar, H.M., R.T. Wood and J. Byrd, *Process and Technology Readiness Assessment for Implementing Concurrent Engineering*, Proceedings of the Second Annual International Symposium of the National Council on Systems Engineering, Seattle, 1992
7. Kota, S. and C.L. Lee, *A Functional Framework for Hydraulic Systems Using Abstraction/Decomposition Hierarchies*, Computers in Engineering, Proceedings of the 1990 ASME Conference, vol. 1, 1990
8. Mäntylä, M., A Modeling System for Top-Down Design of Assembled Products, *IBM Journal of Research and Development*, vol. 34, no. 5, 1990.
9. Nevins, J.L. and D.E. Whitney, *Concurrent Design of Products & Processes, a Strategy for the Next Generation in Manufacturing*, McGraw-Hill Publishing Company, 1989
10. Salzberg, S. and M. Watkins, Managing Information for Concurrent Engineering: Challenges and Barriers, *Research in Engineering Design* 2, pp.35-52, Springer Verlag, 1990
11. Yoshikawa, H. and D. Gossard, *Intelligent CAD*, Part I, Proceedings of the IFIP TC 5/WG 5.2 Workshop on Intelligent CAD, North-Holland, 1989
12. Yoshikawa, H. and D. Gossard, *Intelligent CAD*, Part II, Proceedings of the IFIP TC 5/WG 5.2 Workshop on Intelligent CAD, North-Holland, 1989

17

Human resource management in customer-driven manufacturing

17.1 INTRODUCTION

In classic economic theory, resource management is defined as the challenge of providing the elementary production factors "capital equipment" and "materials" to be combined in the production process. More recent theory identifies "information" as another elementary production factor. This was discussed in detail in Chapter 5, where the theoretical framework of this book was introduced. Within this framework, the resource view was positioned parallel to both the work-flow view and the decision view. Because of the importance of human resources in customer-driven manufacturing (as explained in Chapter 3), this chapter concentrates on human resources.

The focus of resource considerations has always been to calculate the amount of resources per product consumed in the production process. Various types of methods are available with which the amount of work spent in the production process can be determined, the depreciation of the machines can be calculated and the consumed materials, energy and work can be accounted for. This emphasis on resource consumption leads to the understanding that resource management has mainly to do with the optimisation of resource consumption.

In customer-driven production systems, this exclusive focus on consumption is no longer sufficient. Because the product requirements are usually not fixed and known, the main challenge of resource management is not to calculate but to anticipate the required capabilities and capacities needed to fulfil a range of specific customer requirements. The primary problem in these cases is not optimum consumption of production factors but fast and reliable availability of production factors.

The various requirements for resource management in make-to-stock and customer-driven production systems are summarised in the following table:

type of resource	make-to-stock systems	customer-driven systems
"worker"	specialists	generalists
"equipment"	one-purpose	multi-purpose
"material"	broad scope	limited scope
"information"	directive	instructive

Table 17.1 Requirements for various types of resources in customer driven systems

With reference to human work, classical product-oriented make-to-stock systems usually try to exploit the learning curve by organising production as a sequence of simple repetitive tasks. These systems strive to find specialisation and the optimum balance between skills and predefined tasks. This approach is not useable in customer-driven systems, because these are usually not based on the batch sizes needed to exploit the learning curve. Faced with different individual customer requirements, these systems must rely on skills, experience and creativity rather than on predefined routines. As a result, customer-driven systems usually require generalists as human resource in order to cope with different customer requirements.

Note that this difference is also valid for the resource "equipment": Customer-driven production systems require a broad scope of multi-purpose equipment as tools to handle a wide range of products or services.

The comparison with regard to materials shows yet another difference: product-oriented systems are usually not limited to a specific range of materials used in the production process. Today's automotive industry, for example, handles a wide variety of materials, parts and components. Here the functionality of the material used is limited, not the variety of the material. This is different in resource-oriented systems: these systems are limited to a specific material such as sheet metal, plastic or wood. The system has a lot of experience and expertise with reference to this specific material, and a wide variety of products can be developed based on the same material. For instance: a carpenter can make different products from wood, but he will not be prepared to realise the same functionality in rubber or plastic. He will usually tend to make products based on wood.

With regard to "information", another important difference is seen between the two types of systems: in product-oriented systems information is normally used to control the production process. The main function of information is therefore directing the production process. In resource-oriented systems the function of information is broader. In addition to the control function, other information functions are instructing what to do and how to do it and creating awareness and mutual understanding. This places different requirements on information systems as a resource and function in the production process.

17.2 BASIC RESOURCE MANAGEMENT REQUIREMENTS

The examples above show that resource management - especially in resource-oriented systems - requires more than calculating and planning the consumption of resources. We will therefore attempt to discuss some criteria for resource management requirements according to the categories above, but now in the reverse order.

Starting with "information" as the resource category, information should be instructive in customer-order-driven systems and meaningful to the people involved. Although this would appear obvious, in current practice the distribution of information is still treated very restrictively and from the viewpoint of upper management. If information is to be a flexible resource, information management in resource-oriented systems should be organised according to pull principles. Information should be provided on demand, at the requested time, in the requested way, at the requested place.

Properties of resources

In addition to a distinction of resources according to the elementary production function, resources can be distinguished according to their properties for acquiring resources. Here the following distinctions can be made in describing four different types of resources:

a. *Durable and non-durable resources*

A durable resource represents a pool of services available for a certain range of capabilities usually having a decreasing performance potential in the course of time. Examples are tools, machines and buildings. Non-durable resources are consumed in a production process as input. Typical examples here are raw materials and energy. These resources must be provided from stock or by continuous supply.

b. *Human and non-human resources*

Human resources can also be described as a pool of services available for a certain range of capabilities. The main difference between human resources and non-human resources is that their performance structure changes with time. This is illustrated in Figure 17.1.

This picture shows[1] that the performance structure of a human being changes with age. The maximum level of physical power is reached in the person's twenties. With increasing age the physical performance limit will decrease, especially from the fifties onwards.

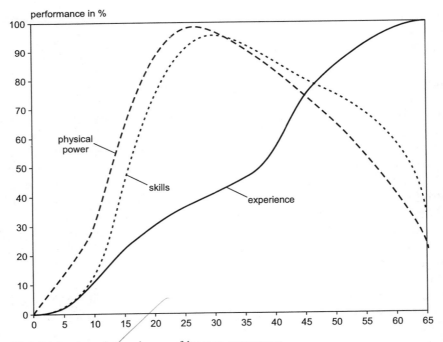

Fig. 17.1 Age dependency of human resources

---

1   We do not claim that these curves represent the true process of performance development in the sense of an empirically-proven graph. The picture is simply an illustration of findings from medicine and geriatry.

Skills, however, need more time to mature and will typically reach their maximum in the person's thirties. This performance level will be maintained for a period of time before it gradually decreases in the person's fifties. The third curve represents experience. Experience needs an extremely long production time and will reach the maximum at the end of the person's professional life. One important consequence of this picture is that the common misconception that older employees have a lower performance is only true for work involving primarily the physical power component of human resources.

The second significant property of human resources - particularly as compared to machines - is that human performance will grow as the resource is used. While the performance limit of machines does not increase but rather decreases with use, a growing performance over lifetime is a fundamental human characteristic. Moreover, continuous stress and strain slightly beyond the actual performance limit of a human being is a good provision for growing performance[2].

Human resources must (can) be hired, trained and motivated to be made available as resources.

c. *Tangible and intangible resources*
Tangible resources are resources that can be touched or grasped while intangible resources cannot. Examples of tangible resources are fixed assets, money, materials and natural resources. Examples of intangible resources are ideas, skills, experience, organisation and procedures. Tangible resources are usually directly countable (quantifiable), while intangible resources must be represented by a description.

In so far as tangible and intangible resources are goods or non-personal services, they must (can) be bought, shipped and stored. Intangible resources, offered as personal services, must (can) also be bought but cannot be shipped or stored separated from the carrier. All other intangible resources cannot be bought but must be produced and maintained.

d. *Visible and invisible resources*
Visible resources are resources that can be observed in reality;

---

[2] At least for the increasing part of the curve; beyond the peak stress and strain usually accelerates the rate of descent.

invisible resources are real but not observable. This distinction appears similar to the previous distinction; a significant difference however is that invisible resources cannot be represented by a formal description.

The concept of *tacit* knowledge[1] is a good example of resources of this type: *tacit* knowledge designates that part of (human) knowledge which is only implicitly available or embedded in a skill, but which cannot be made fully explicit[3].

Invisible resources can neither be bought nor produced directly; they must grow. You can only try to establish framework conditions within which they can grow.

17.3 WORK SYSTEMS AS A MEANINGFUL SYNTHESIS OF RESOURCES

The synthesis of resource elements is called the design of a work system. A work system is defined as a purposefully-organised system in which employees and other resources work together in a work process, on distinct work stations and in a work environment under the conditions of this work system [2]. Descriptive elements of a work system are: the duty of the work system, the input for a work system such as material, energy and information, the output resulting from the work-process in quantity and quality, the acting elements in the work system in kind, quantity and location (e.g. humans and machines), and the environment and context of the work system.

In addition to the description of elements, a work system is determined by the system level. According to H. Ulrich we can distinguish between three types of systems: "Dead", "Living non-human" and "Living human" systems (see Figure 17.2).

Each of these types covers different levels of the system's perspective. The perspective level of "dead" systems is the material they consist of. Specifying the components of such a system will render an adequate description. With "Living non-human" systems, the perspective of how they function arises as an additional level. Here a description of the components alone is not sufficient - the relationships in the system

---

3   Another known example for the distinction between visible and invisible is the distinction between data and information. You can describe and store the data part of information, but not the information itself.

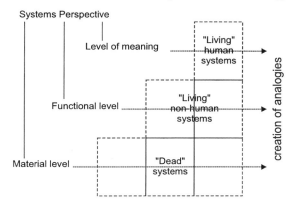

Fig. 17.2 System levels

and the behaviour of the system are necessary integral aspects. "Living human" systems require the additional perspective of the meaning constituting the system. Without meaning "Living human" systems are incomplete and cannot work.

The actions of human systems are not limited by the built-in functionality; they are able to transcend the given functionality by a self-referencing creative process[4]. Human systems, equipped with meaning, have very low external control requirements because they are able to create solutions independently for certain problem specifications.

system types	behaviour control interface
"dead" systems	passive operations
"living non-human" systems	reactive target values
"living human" systems	active problems

Table 17.2 Characteristics of the three system types

---

[4] The theoretical concept behind this is called "Autopoiesis".

17.4 CHALLENGES FOR RESOURCE MANAGEMENT OF RESOURCE-ORIENTED SYSTEMS

After this theoretical excursion on resources, the obvious question is what the consequences of these considerations are for the (re-)design of resource-oriented systems. Our current understanding of customer-driven systems is not yet sufficient for an elaborate answer to this question. A considerable amount of research must be done. But we can give some guidelines to pinpoint general postulates for resource management in customer-driven systems. These postulates can be summarised as follows:

a. Evolutionary Human Resource Management
b. Complex Resource Units
c. Holistic Work Design
d. Healthy Informal Organisations
e. Concurrent Mental Modelling

a. *Evolutionary Human Resource Management*
The challenge of Evolutionary Human Resource Management (HRM) highlights the view of human resources in customer-driven systems as an organic asset. HRM is not like material management, and counting the heads of employees is therefore not sufficient. The number of employees is a figure commonly used to determine the capacity of a company, but the internal structure of this figure determines the company's ability to produce, reproduce, transfer and enhance the embedded living knowledge and experience. Because internal knowledge and experience transfer loops between generations of employees in a company are often long-lasting processes, careful management activities are required to support this process.

b. *Complex Resource Units*
The challenge of complex resource units simply emphasises that the resource units in customer-driven systems should provide a high level of complexity in terms of capabilities. The one-task-one-employee principle in Adam Smith's pin factory is not adequate for customer-driven systems. Adam Smith's principle is based on standardisation and specialisation; in customer-driven systems people are faced with non-standard, fuzzy situations. This requires a complexity of capabilities that a single employee cannot offer. As a result, group work and the configuration of stable work teams is an essential

subject of HRM in customer-driven systems. The postulate of group work and work teams as resource units involves numerous questions about optimal group size, mix and structure of groups.

c. *Holistic Work Design*

The challenge of a holistic work design reflects that customer-driven systems are experience-driven and that this experience must be acquired and maintained during the working life. Because customer-driven production often requires a simultaneous product and process design order, it is extremely important that employees capture the wholeness of this process. Capturing this wholeness requires involvement in a holistic loop in order to generate experience.

If someone has experience, this experience can be applied to manage situations. An existing level of experience can therefore be used to generate new experiences. These new experiences can be gained either by direct actions (trying something new) or by exchanging experiences through interaction with other people. Naturally, these new experiences will raise the level of existing experience, thus completing the loop.

d. *Healthy Informal Organisations*

The challenge of healthy informal organisations highlights a different view of organisations. In the past, management theories stressed the role of formal organisations as a means for structuring resources, assigning responsibilities and managing control. Although the existence and importance of informal organisations has been well-known since the Hawthorne experiments, main emphasis was placed on reducing informal organisations to supplementary functions such as dealing with exceptions. In customer-driven systems it is recommended that a healthy and well-developed informal organisation be nurtured. Informal organisations, for example, are extremely useful in building trust between employees; formal organisations may resolve problems between employees by elaborated procedures, but they will never be able to establish trust. Trust is a very important resource in business to facilitate the readiness for concurrent problem solving. The example of the world-wide stock exchange business is an illustrative example of how mutual trust keeps a business working. The support of informal organisations is therefore a useful effort in customer-driven systems. The network and broad-band communication technologies of today are useful means for supporting the development of healthy informal organisations by offering complex interaction facilities as a network rather than as a hierarchy.

e. *Concurrent Mental Modelling*
The challenge of concurrent mental modelling is a very new subject in R&D. The importance of this subject for customer-driven systems is based on the fundamental differences in product and process engineering between customer-driven production and mass production.

In mass production, product development, process development and manufacturing are organised as sequential and separate steps. Process engineering starts when the product development is finalised, and production starts when both engineering tasks have been completed. But this is an ideal model. Industrial reality shows that after starting series production, a great deal of redesign effort is needed to gradually improve products and processes to the desired level. The significant difference between series production and customer-driven production is therefore not linear versus cyclic processes: the most significant difference is that in series production engineers are able to work on real products and processes, while engineering in customer-driven production must to a wide extent deal with virtual product and process development. In customer-driven systems one cannot build full-scale products to evaluate the result; imagined products and processes must be evaluated. To a significant degree, engineering in customer-driven systems is based on virtual models in people's heads rather than on physical models that can be touched and grasped. The risk of misunderstandings and divergent perspectives is obviously greater in virtual modelling. This is the reason why concurrent mental modelling should be advocated as an integral part of engineering in customer-driven production. To understand this, we should refer to some elements of the work process in engineering.

Engineering is not simply placing visions in a void. Engineering basically designates a process for aligning visions with existing knowledge as enabling conditions and constraints. The knowledge to be considered in engineering is usually complex and incorporates product, process, market and financial aspects. As the discussion above on Tacit knowledge illustrates, this knowledge is only partly documented in a company; much of this knowledge is only implicitly available and must be activated in the engineering process. Engineering should

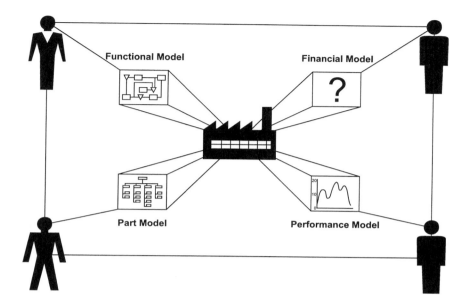

Fig. 17.3 Model integration as a process of concurrent mental modelling

therefore be understood to be not an individual but a collective, cooperative process.

The concept of Concurrent Engineering reflects these facts and considers engineering from a round-table point of view, where people from different disciplines sit together to simultaneously define products and processes. This means that the individual visions and knowledge must be amalgamated into shared visions and knowledge.

Shared vision is needed as a common guideline in support of consistency of the results, and common knowledge is necessary to integrate all constraints. How can this amalgamation take place if the whole is split up into different views with no defined overlap? Figure 17.3 represents a situation of this type, in which a product is envisaged in different models from different views by different actors.

In customer-order-driven systems, neither homogeneity of professions nor long-term processes of mutual accommodation are satisfying solutions. Here solutions must be applied that go beyond the scope of individual professions, and time is too short to leave mutual understanding to a process of natural growth. Supporting integration models requiring an extremely high level of education in order to understand and work with them is not sufficient. What is needed is not only the integration of routines and data but also methods and procedures bridging the gap between various views of dismayed persons. The concept of concurrent mental modelling therefore prescribes active support of mutual understanding as a necessary complement to logical and formal integration.

This is a field in which simulation and animation can play a major role in the future. The benefit of simulation and animation in this context is to experience new or unfamiliar relations in an impressive way. It is not simulation's ability to accurately predict; it is the intuitive experiencing of fundamental principles and consequences of new views that it offers that is the power of simulation in customer-driven systems. In addition, management science has developed powerful methods and techniques (e.g. role playing and games) to facilitate mutual understanding and to break through solidified interaction rituals. These methods and techniques are extremely useful for concurrent mental modelling, but they have not yet been applied and adopted to the engineering situation in customer-driven production systems. Further research and development for dedicated solutions is required in this respect.

REFERENCES

1. Polanyi M., *The Tacit Dimension*, Garden City N.Y., 1966
2. See DIN 33400 (1975)
3. Maturana H., Varela J., *Autopoiesis and cognition - The realization of the living*, (1980), D. Reidel Publishing Company, Dordrecht, Holland.

Part D
ORGANISATION AND DECISION MAKING

18

Organisation and decision in customer driven manuf...

18.1 INTRODUCTION

The main features of customer driven manufacturing systems were presented in Part A of this book. A theoretical framework elaborating on the Walrasian model was presented in Chapter 5. In this chapter, we will focus on the organisational and decisional aspects.

We will first discuss in section 18.2 the requirements which follow from the competitiveness of customer driven manufacturing systems. Then we will introduce in subsequent sections a generic model which is suitable for describing organisational and decisional structures, the GRAI model. Finally, we will show the instantiation of the GRAI generic model in the domain of customer driven production and engineering.

18.2 REQUIREMENTS FOR COMPETITIVE CUSTOMER DRIVEN MANUFACTURING SYSTEMS

Related to customer driven manufacturing (including customer driven engineering), the new challenge here is to be able to integrate various types of activity such as intellectual activity (engineering) and physical activity (manufacturing or assembly) in the same customer driven process. *Time* appears to be a key performance indicator of customer driven manufacturing. The main objectives for such a manufacturing system, and also in standard production, are to reduce time to market, to shorten manufacturing lead times and to increase reactivity and flexibility. These points are all dependent on time. Therefore, when talking about integration, the main feature we need to take into account is time.

Designing manufacturing systems involves several aspects. First of all, the physical production environment should be considered for redesign, aiming at *physical integration*. This aspect was discussed in Part C of this book, in particular in Chapter 15. Developments such as dedicated product flow lines, automatic storage systems and robotics help to improve the physical flows in terms of speed, reliability and accuracy. However, a high level of automation may lead to decreasing flexibility, which is not acceptable in customer driven manufacturing. New organisational principles such as teamwork, in which various skills work together, may help in improving the integration level of the physical system. This point was discussed in Part C, Chapter 17.

Secondly, the integration of physical flows cannot be realised without information systems. For customer driven engineering activities especially (cf. Chapter 14), information management is essential. This second aspect may be called *informational integration*.

The third aspect to be mentioned here is *decisional integration*. This aspect is closely related to organisation, which is one of the key aspects to be considered in customer driven manufacturing. Whatever the level of integration reached for the physical and the informational aspect, the performance of the manufacturing system will depend also on the degree of integration of the decisional system.

Finally, the overall competitiveness of manufacturing systems will not only depend on the level of integration for each aspect individually, but also on the level of integration between these aspects. The performance of the manufacturing system requires integration between the physical, the informational and the decisional aspects.

We need a model with sufficient genericality in order to take into account all the particularities of the various manufacturing systems and the integration features. We propose to use the GRAI model.

18.3 THE GRAI MODEL

18.3.1 Basic concepts

The generic GRAI model [Doumeingts 84] was developed in the early eighties in order to allow production management systems to be described and to facilitate the choice of Computer Aided Production Management Packages. A specific description of a particular production management system is called a GRAI model. The various phases passed through in the design of a GRAI model and in the choice of packages was called the GRAI method. We will present two formalisms used in this method, the GRAI grid and the GRAI nets:

- The GRAI grid represents the global model of the decisional structure. It plays an important role in defining the integration. We call this model the macro-model of the decision structure.
- The GRAI nets describe the decision centres. We call this model the micro-model of the decision structure. This micro-model is related to resources, particularly human resources, which are one of the key components of the organisational structure.

Development of the generic GRAI model was based on the problems encountered by the GRAI researchers when they attempted to design models of production management systems. The basic concepts of GRAI models are supported by several theories: systems theory [Le Moigne 90], organisational theory [Mintzberg 84], hierarchical theory [Mezarovic 70]. The GRAI model at conceptual level is presented in figure 18.1. It allows the production system to be split into three subsystems: the decisional system, the informational system and the physical system.

The *physical system* consists of all the activities related to realisation of the product. It covers all the activities which give added value to the product. In customer driven manufacturing it goes from the engineering activities (product definition) to product delivery.

The *informational system* is in charge of collecting, memorising and storing information on the physical and the decisional activities, as well as providing these activities with information.

The *decisional system* is composed of the control decisions aimed at managing the physical system. The decisional system is split into two domains: the periodic domain and the event-driven domain.

The periodic domain, which covers the strategic and tactical level and a part of the operational level, is activated by periodic decisions. For example, each month the master scheduling plan is calculated. The event-driven domain corresponds to the lowest level of the decisional structure: this domain is activated by events. For example, when a part is finished, ready for being machined on a work station (an event), the transportation system is activated to move the part to the next work station. This chapter focusses on the periodic domain.

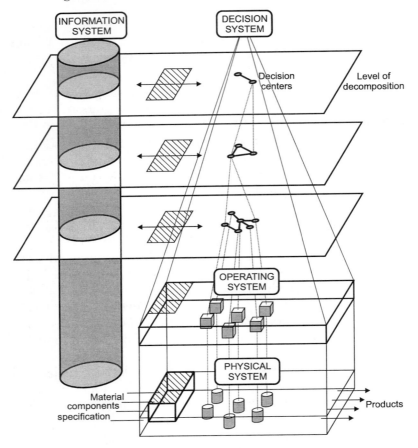

Fig. 18.1 The GRAI conceptual model of production systems

Being a conceptual model for production management systems, the GRAI method was designed according to production management concepts. The role of the production management system is formally and progressively to prepare, level by level, through a decisional process, the conditions for the physical realisation of the production objectives. This must be done coherently, with an increase in 'executability', and it ends in the release to the physical system of work orders and workable operational instructions (as late as possible) [Marcotte 90].

When considering the physical activities, represented in figure 18.2, the production management system (PMS) can be characterised as

follows. The PMS coordinates between various levels an
in time:

- the processed flows (raw material, components, prc
- resource availability (human and technical resources).

According to the conceptual GRAI model (figure 18.1), we consider two decomposition criteria to structure the PMS for the periodic domain:

- functional decomposition (horizontal decomposition criteria),
- temporal decomposition (vertical decomposition criteria).

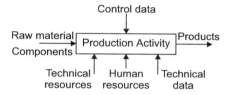

Fig. 18.2   Production activities according to the IDEF formalism

18.3.2 Decomposition criteria

Horizontal decomposition

Horizontal (functional) decomposition is based on the type of management decision. From the PMS definition, we consider three basic types of decision:

- decisions on resources or decisions on capacity,
- decisions on flows,
- decisions on synchronisation (planning decisions).

Note that these three types correspond to the three views of manufacturing described in Chapter 5 while elaborating the Walrasian model.

The decisions on resources are concerned with the availability of Resources (R) in Time (T). The term "resources" refers to all entities which support activity realisation, such as machines, human resources, transport vehicles (for a transportation activity), etc. Together, these decisions constitute the organisational function which manages capacity evolution over time. The name of the function is "to manage the resources".

The decisions on flows (information, raw material, parts, components, finished products) are concerned with Products (P) in Time (T). Again, these decisions together constitute an organisational function. The name of the function is "to manage the products".

The meaning of "product" depends on the context of the physical process. In an engineering department, for example, the element which is transformed is a piece of information (the product definition). This information represents the "product", i.e. the result of the engineering activity.

The decisions on synchronisation (planning) are concerned with the Products (P) and the Resources (R) in Time (T). For example, at the short-term level, it is to ensure that purchased materials and the machine that is going to process these materials will be available at the same time. In the medium term, it is to synchronise material requirements and capacity requirements by planning manufacturing activities. Analogous to the above discussion, the synchronisation decisions together constitute a third organisational function. The name of the function is "to plan".

Vertical decomposition
Vertical decomposition is aimed at identifying the levels of control. A well-known rule is to identify three basic levels of control: the strategic level, the tactical level and the operational level. The higher levels are in charge of defining the overall production features in terms of capacities, purchasing policies, inventory management principles, etc. At this level of control, the information detail level is low (aggregated information) and the time scale is large. The lower levels are in charge of controlling a specific part of the process. The level of detail is more important and the time scale is reduced. This is short-term management.

Therefore, the criterion of decomposition and structuring of decisions is based on *time*, with two related concepts - the decision horizon and the decision period. The *decision horizon* refers to the length of time taken into account in a particular decision (the horizon usually depends on production and procurement lead times). The *decision period* relates to the interval of time after which the validity of the decision is reconsidered [Doumeingts 84]. It is determined by the frequency of the relevant variation in the information used to take the decision. For the upper levels, the period depends mainly on the evolution of the environment and for the lower levels mainly on the behaviour of the physical system (events and perturbation occurrence frequency).

18.3.3 The GRAI grid and the GRAI nets

In accordance with this decisional approach and the decomposition criteria presented above we can build the GRAI grid, which is representative of the structure of the PMS (figure 18.3). The columns are defined according to the type of decision, the levels identified according to the horizon and period of decision-making. The intersection between a column and a level conceptually defines a decision centre.

This GRAI grid contains two other columns, which represent the informational links with the outside of the system (external information) and the information proper to the production system (internal information).

Overall coherence in this decomposition is shown through the links between the decision centres. Two types of link are considered:

- the decisional link,
- the informational link.

A *decisional link* defines the coordination and synchronisation inside the grid. A decisional link is a decision frame which includes objectives,

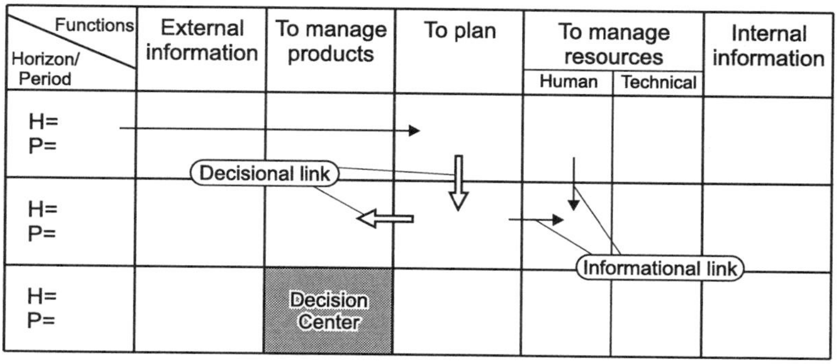

Functions Horizon/ Period	External information	To manage products	To plan	To manage resources		Internal information
				Human	Technical	
H= P=						
H= P=		(Decisional link)				
H= P=		Decision Center			(Informational link)	

Fig. 18.3 The basic GRAI grid

230 *Organisation and decision making in customer driven manufacturing*

decision variables, constraints and criteria. A decision frame "controls" or "sets the frame" for another decision centre located at the same level or at a lower level. A decision link is represented by a double arrow or a large arrow.

An *informational link* represents the main information (information which does not specify objectives, decision variables, constraints and criteria) exchanged between two decision centres or between a decision centre and one of the information columns.

The GRAI nets give the structure of the various activities in each decision centre identified in the GRAI grid. By using GRAI nets, the result of one discrete management activity can be connected with the support of another management activity. GRAI nets allow four decisional elements to be identified:

- the nature of the decision or transformation (activity name),
- the initial conditions (main input of an activity),
- supporting elements (information, objectives, decision variables, constraints, resources),
- results (outcome of an activity).

The decisional links are explicitly detailed through the description of the objectives and decision variables employed in decision-making. The representation is illustrated in figure 18.4.
In the GRAI nets, the information related to the decision allows the study of some organisational features. Actually, the organisation is

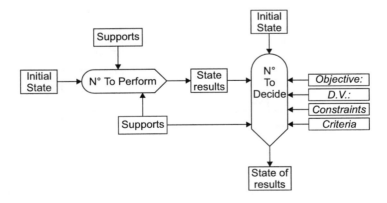

Fig. 18.4 GRAI nets formalism

based on responsibility and authority structures, which relate respectively to objectives and decision variables in the GRAI nets.

18.4 MODELLING ENGINEERING ACTIVITIES USING THE GRAI MODEL

In the typology of customer driven manufacturing situations, presented in Chapter 6, there is one type which is intuitively the most extreme customer driven case, viz. the case of resource-oriented engineer-to-order manufacturing. In this book, the shipyard presented in Chapter 12 is an excellent example.

Before proceeding with a description of this type of industry in terms of decision structures, a short comment on the organisational structures encountered should be given. Two main types of organisational solution can be found in practice for resource-oriented engineer-to-order manufacturing.

The first solution is a multi-project organisation. An organisation of this kind proceeds by grouping all the resources required for a particular project in a project organisation.

The second type of organisation proceeds by breaking down the project (the customer order) into various suborders, distributed to the specific departments involved. Each department corresponds to a specific skill, e.g. an engineering department, a manufacturing department or an assembly department. Each of these departments is then involved in several parallel subprojects. In this case, the main organisational structure is not the project structure, but the departmental structure.

The type of organisation chosen depends mainly on the amount of the company resources required for each project. In the second type of organisation, each project represents a low workload for each specific function or department. In the remainder of this chapter we will concentrate on the second type of organisation.

18.4.1 The specific nature of engineering activities

Engineering activities cover various phases such as product design and process design.
If we describe the "physical system" of the engineering activity using the same concepts given in figure 18.2, we obtain figure 18.5. The main

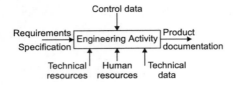

Fig. 18.5 Engineering activity according to the IDEF formalism

difference comes from the nature of the object being transformed. Manufacturing activities transform physical objects while engineering activities transform information. The nature of these activities is quite different, also from a management point of view. The following points can be highlighted with respect to engineering activities:

- they transform information,
- their duration is uncertain,
- their results are not known *a priori*,
- their evaluation depends on the subsequent activities of the process (for example, the quality of process planning will be evaluated during the manufacture of the product).

The other activities (manufacturing, assembly, delivery, etc.) are applied to physical objects. These activities can be characterised as follows:

- they transform physical objects,
- their duration is known (when the activity is defined) *a priori*,
- their result is known (when the activity is defined) *a priori*,
- their evaluation does not depend on the other process activities.

It is clear that engineering activities increase the uncertainty related to customer driven processes. This uncertainty needs to be managed. The approach to be taken here is to manage information flows as a "product flow".

18.4.2 The GRAI modelling approach for engineering management

The objective of the function "to manage products" can be summarised as "to provide the right product in the requested quality and quantity at the right time". This objective obviously covers product supply, product transportation (flows) and product storage.

Now, in the case of engineering activities we have to consider information as the product. However, the objective of the function "to manage information" is the same: to provide the right information in the requested quality and quantity at the right time. The basic elements of this "to manage information" function are Information (I) and Time (T).

For instance, when discussing matters with people working in the engineering department, we find their main requirements are:

- to have rules, approaches, tools to help and speed up the definition of customer requirements
- to have tools (DSS, Expert Systems, etc.) to extract and formalise experiences
- to store information in a way which allows easy access for re-use (structured databases, classification criteria, etc).

So, for the management of these engineering activities it is very important to take into account information management. This concept is fully related to product management in the management of manufacturing activities [Marcotte 92].

With regard to these previous remarks we can build the GRAI grid specifically for an engineering department in which the function related to flow management is called "Information Management". An example is presented in figure 18.6. This grid represents the decisional structure for the management of the engineering department.

18.5 CDM SYSTEMS: A REFERENCE MODEL FOR MANAGEMENT STRUCTURE, BASED ON GRAI MODEL

In the environment of customer driven manufacturing, it is important that the production management system should cover the overall

Functions \ Horizon/Period	External information	To manage products	To plan	To manage resources	Internal information
H=1 year P=1 month	Forecasts Orders Commercial policy Production plan	- technology survey - knowledge acquisition - info mgt. policy	Yearly activity program (new development + customer related activity)	Definition of the requirements in human and technical resources	
H=2 months P=1 week	Orders Quotation requests Customer requirements	To program technical data updating	Planning of the engineering activity	Human resources sharing out	
H=2 weeks P=1 day	Sub-contractors Customer Orders Quotation	To check info. availability To specify info. requirements	Scheduling	Task allocation	
Real Time		To store information, to build up experience	Dispatching	To design	Follow up

Fig. 18.6 Example of a GRAI grid for engineering activities

process and that there be a real integration of the various activities. In the following illustration the types of activity to be considered are Engineering, Parts Manufacturing, and Assembly.

In the case of a pure project organisation, the integration of the various activities is achieved through the principles of teamwork. Then, for project management, one grid can be used to structure the production management system of this dedicated unit. The main difficulties occur when one unit has to deal with several projects at the same time. In the following, we focus on this last case.

18.5.1 Management structure decomposition: multi-grid organisation

At strategic level, management is concerned with the overall manufacturing process. At lower levels, management actions are applied to different production units or departments according to the organisation of the production system. For instance, an engineering department, machining unit, assembly unit can be the various production units. For each production unit there is a grid representing the production management structure of this production unit.

The management structure is split up according to the production unit organisation being considered. The overall objectives are split up in an inclusive way, because of the specialisation of the production activities. Then, from one grid at the level of top management, several

CDM systems: a reference model for management structure 235

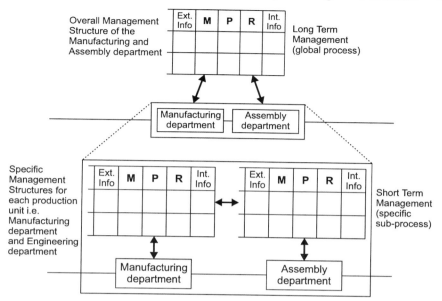

Fig. 18.7 Overall management structure covering the manufacturing and assembly process, and the specific management structure for the manufacturing department and for the assembly department

grids may appear for the management of the various production activities (figure 18.7). In this figure, the letters are representative of the three basic functions: to manage products (M), to plan (P) and to manage the resources (R).

18.5.2 Interdepartmental integration

When one department is studied, activities from other departments or units influence the management activities of the department being studied. It illustrates the integration requirements presented above. For instance, the engineering department will influence the management of the manufacturing unit. When studying the manufacturing department only, it is important to take into account these interrelationships, so a column called "To manage engineering" is added (figure 18.8).

(The arrows in the following figures are inserted as examples and characterise only the presence of certain links and not their type, i.e.

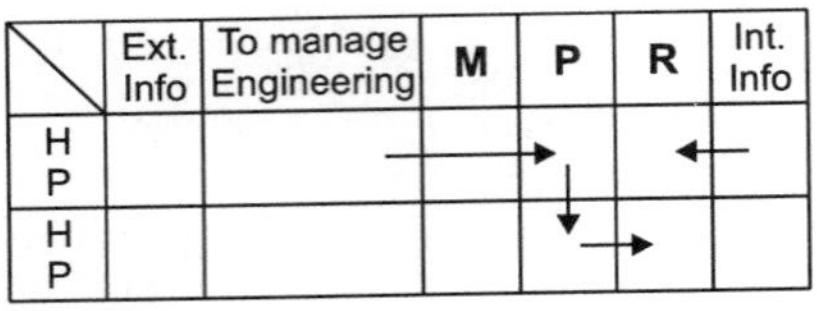

	Ext. Info	To manage Engineering	M	P	R	Int. Info
H P						
H P						

Fig. 18.8 Management structure of the manufacturing department, and its specific interrelationship with the engineering department

decisional or informational. The links inside the management structure of each department management structure are not represented).

In the same way, when we study the engineering department, if we want to show the relationships with the manufacturing department it is possible to add a column "To manage manufacturing" (figure 18.9).

Thus, the GRAI model allows integration, at least decisional integration, to be taken into account when specifying the coordination rules and, partially, informational integration when defining the requirements for information exchange in the production management system. Physical integration is more concerned with the description and analysis of the production process.

Various kinds of organisation may be investigated according to the type of management system. In the following example (see appendix), we intend to present two types of organisation. They are based on the degree of engineering that is customer order driven. Production in that case is more or less project-oriented.

In one case (example 1), the degree of customisation is very important and multi-project planning (starting from customer order acceptance) covers the three basic steps: engineering, manufacturing and assembly. A structure of this kind can be employed for shipbuilding or civil engineering, for instance (cf. Chapter 12).

In the second example, the customer order driven engineering activities are less important and are in general mainly performed during

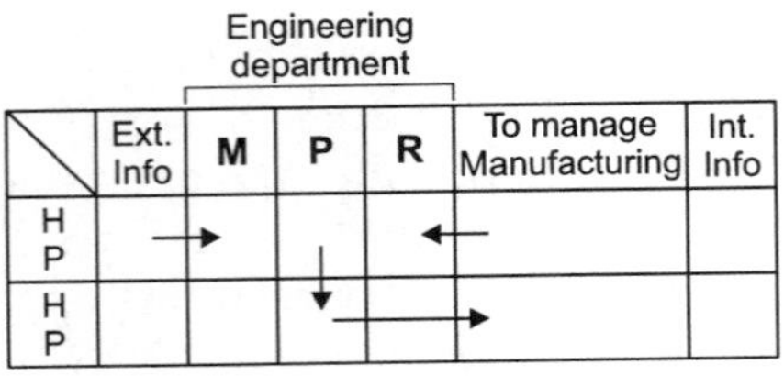

	Ext. Info	M	P	R	To manage Manufacturing	Int. Info
H P						
H P						

Fig. 18.9 Management structure of the engineering department and its specific interrelationship with the manufacturing department

the quotation phase. Then, multi-project planning is more concerned with the manufacturing and assembly phases. This type of structure can be employed for industrial robotics or industrial electrical networks, for instance (cf. Chapters 9 and 10).

18.6 CONCLUSION

Nowadays, production management can no longer be reduced to a purely technical approach. The evolution of production systems and the new requirements for integration lead to the development of new approaches in which the various aspects of production systems are taken into account in an overall modelling framework. The GRAI model can provide some answers to the requirements.

Regarding this approach to customer driven manufacturing, it is now necessary to develop specific techniques and tools for the various functionalities we can identify, e.g. multi-project planning, engineering scheduling, information management, and human resource management. In the following Chapters, some potential solutions are presented to support these functionalities.

REFERENCES

1. Doumeingts G., "Méthode de conception des systèmes de Productique", *Thèse d'état en Automatique*, Laboratoire GRAI, Université de Bordeaux I, 1984.
2. Le Moigne J.L., *La modèlaisation des Systèmes complexes*, Presses Universitaires de France, 1990.
3. Marcotte F., "Organizational / Decisional view", *FOF ESPRIT BRA 3143*, GRAI University of Bordeaux I, June 1990.
4. Marcotte F., Doumeingts G. and N. Ould Wane, *Management of Engineering activities and Manufacturing Activities in One of a Kind Production, using the GRAI approach*, GRAI University of Bordeaux I, June 1992.
5. Mesarovic M.D., D. Macko and Y. Takahara, *Theory of Hierarchical, Multilevel Systems*, Academic Press, New York 1970.
6. Mintzberg H., *The structuring of organizations: a synthesis of the research*, Prentice-Hall Inc, 1982
7. Rolstadas A., "Design of the Conceptual Model", *FOF ESPRIT BRA 3143*, SINTEF NTH, November 1990.

APPENDIX I
Example I
Shipbuilding

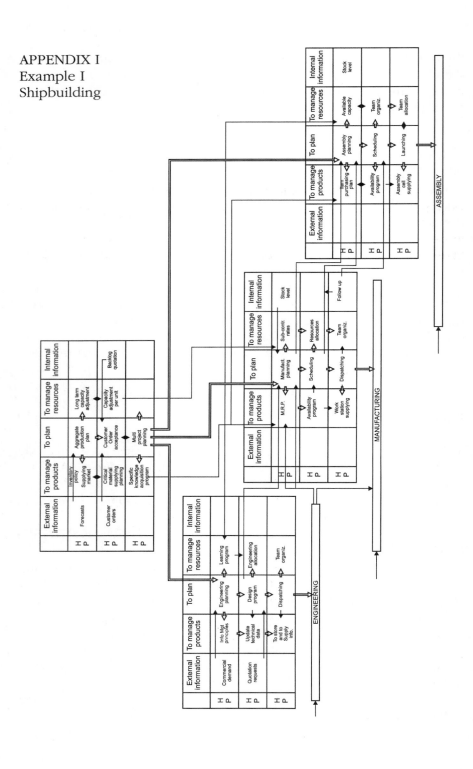

APPENDIX II
Example II

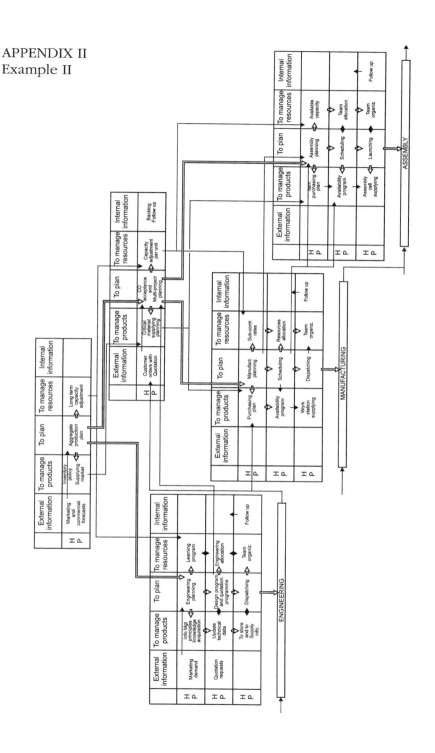

19

Production control in workflow-oriented make-to-order firms

19.1 INTRODUCTION

Customer driven manufacturing is usually associated with capital goods. This book takes a wider perspective: customer driven manufacturing occurs in many branches of industry. A typology of customer driven manufacturing was presented in Part A, Chapter 6. Perhaps the most remarkable point was the fact that many workflow-oriented industries are driven by customer orders. Examples of these factories occur in industries such as printing, textile, paint or packaging materials. Most notably, traditional make-to-stock products such as steel or paper are becoming increasingly customer order driven in the higher quality product ranges. Therefore, a case from the fine-paper-making industry was included in Part B (Chapter 11).

The concepts for production control in this type of industry will be presented in the present chapter. In particular we will concentrate on make-to-order rather than on assemble-to-order (which will be discussed in the next chapter) or engineer-to-order (which is discussed in Chapter 21).

In a way, make-to-order production control is easier than engineer-to-order or assemble-to-order. In engineer-to-order companies there is always considerable uncertainty about the nature of the product to be supplied. In assemble-to-order companies the variety of potential customer orders makes forecasting and master production scheduling difficult. In make-to-order companies production control appears to be in the comfortable position of being able to wait for customer orders to arrive and to start working when all uncertainty has been eliminated.

However, real life is not as easy as it may look from a distance. There are several reasons why designing a production control system in workflow-oriented make-to-order companies is difficult:

- Negotiating about due dates is an integral part of production control problems.
- Changeover times usually play an important role in one or more production phases, and this fact makes lead-time estimation difficult.
- The structure of the production system (in terms of the number of production phases for which work orders are to be released) determines the nature of production control.
- The autonomy of different decision functions in production planning and control has to be established.

These issues will be introduced shortly below.

First of all, production control always includes the problem of *quoting due dates* for customer orders.

In make-to-order firms, it is usually necessary to take the backlog of work (i.e. work accepted but not yet finished) into account. The simplest way to do this is by fixing the estimated start date of the order according to the backlog. If the backlog is **B** hours and normal capacity is **C** hours per time unit, it will take at least **B/C** time units to finish the backlog. A more elaborate way of estimating the start date of an order is by assuming a given "normal" workload **WL** in the shop. In that case, the estimated start date of an order is (**B** - **WL**)/**C**. The estimated lead time of the order after being released would be (2 **WL**)/**C**. Therefore, the estimated finish time of the order would be (**B** + **WL**)/**C**. This is illustrated in figure 19 1.

Quoting due dates is especially difficult when combined with a second characteristic of workflow-oriented firms: the fact that considerable *changeover times* occur in important types of equipment. The reason for this is that the lead times of orders become dependent on when machinery will be properly set-up. Where the set-up for a

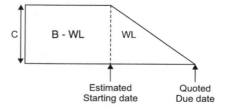

Fig. 19.1 Quotation of due dates

particular type of work is infrequent, an order may have to wait for a long time before it can be processed.

Unfortunately, changeover times are quite common in make-to-order workflow-oriented firms. Actually, nearly all machines in workflow-oriented firms have *some* form of changeover problem and other production constraints, such as a lack of space or widely varying operation times. This fact makes lead-time estimation a hazardous task. One of the ways of dealing with this problem is by enforcing a predetermined changeover schedule for the important machinery. This makes sense in the case of sequence-dependent set-up times. It implies that a distinction is drawn between aggregate production planning (where the changeover sequence and total length is determined) and detailed production planning or factory coordination (where work orders are planned which result in the realisation of customer orders).

Thirdly, it is usually practical to draw a distinction between planning and control in terms of subsequent work orders ("Factory Coordination") and planning and control of individual operations within a work order. A key question here relates to the *structure* of the production system, i.e. which production units are to be distinguished. For example, in the original fine-paper case of Chapter 11 there is only one production unit (phase). However, in the remainder of this chapter we will argue that two phases (production units) are more appropriate.

Quoting due dates is usually a part of factory coordination, whereas realised lead times are the result of production unit control. Factory coordination is often a task of the materials manager (sometimes called "the factory planning department"), whereas production unit control is the task of a production supervisor or foreman.

Factory Coordination and Production Unit Control presuppose each other: you cannot have one without the other. Consequently, the production manager and the materials manager will run into conflicts unless their interaction is carefully designed. We will pay considerable attention to this point in this chapter.

This brings us to the last point to be introduced here, viz. the *autonomy* of people on the shop floor in relationship to the autonomy of materials management. Returning to the issue of changeover sequences, for example, it is important to establish the amount of autonomy of the shop-floor people in determining the set-up sequences of machines.

We will first present in section 19.2 some basics of professional production planning and control — as far as is relevant here. Other chapters in Part D will provide greater insight. Next, in section 19.3 we will discuss the *structure* of the production process. It will be pointed out that it is often appropriate to redesign this structure for production planning and control purposes. Once a structure has been chosen, it is possible to give an outline of various production control functions. This outline is discussed in detail in section 19.4 in connection with the case introduced in Chapter 11. The conclusions of this chapter are presented in section 19.5.

19.2 PRODUCTION CONTROL DESIGN PRINCIPLES

As mentioned in the introduction, several principles of production planning and control will be discussed in this section. More specifically, the following points:

- the role of Aggregate Production Planning, for example, in establishing changeover sequences
- the interdependence of Production and Sales, for example, in quoting due dates for customer orders
- the distinction between factory coordination and production unit control, for example, in determining the number of production units
- coordination between the material and the capacity aspect of production control.

Aggregate Production Planning

It is generally known that volume fluctuations do exist in many markets. Generally speaking, volume fluctuations can be dealt with by:

- varying the available capacity of production facilities;
- varying the available capacity of the workforce through hiring and firing;
- varying utilisation of the available capacity, usually through overtime (increase in working hours per week) or through decreasing working hours per week;
- varying inventories: in customer-driven production, the production backlog takes the role of inventories (production backlog refers to accepted orders not yet in production).

These decisions are often collectively called *aggregate production planning* (or "Production Planning" for short in APICS terminology). The adjective "aggregate" is added in order to stress the fact that these decisions are usually not meant for a particular material item but for a factory or substantial part of a factory.

For the long term, two strategies are possible. At one extreme, there are companies that can hardly afford to adjust their production capacity to these fluctuations. Consequently, these companies suffer from under-utilisation in times of low demand, whereas they can produce at full speed when demand is high. These companies avoid the cost of capacity adaptation, but they will face a loss of market share in good times and cost competition in bad times. At the other extreme are the companies that can adapt their production capacity completely to the volume of sales that can be obtained in the market. These companies face the cost of capacity adaptation, but will generally be cost-competitive in bad times and not lose market share in good times.

In the short run, two other strategies are possible. At one extreme, capacity utilisation can be varied to accommodate production volume to sales volume. This means that the working hours per week of the available manpower is varied. (Subcontracting is another possibility for the short-term expansion of capacity utilisation, but subcontracting is often difficult in customer-driven process-oriented industries). At the other extreme, backlogs of work can be maintained and varied according to the volume of sales. This means that customer order lead times should be variable and that the backlog of orders should be taken into account when quoting due dates for customer orders.

In practice, a combination of the above four methods will be used. Therefore, aggregate production planning cannot be ignored in many cases, and it can provide parameters for other decision functions. In the fine-paper case of Chapter 11, the last method (varying backlogs) is dominant in times when capacity is scarce. In times when abundant capacity is offered, the other methods are also applied.

Production and sales: quoting due dates for customer orders

Generally speaking, negotiating with customers on due dates is indeed a matter of *negotiating*. The method depicted in figure 19.1 is much more in the nature of take-it-or-leave-it. In many cases, this method will yield a quoted customer due date which is too early for the customer. If the due date had been negotiated, it might have been possible to quote a due date which is much later than the one produced by the nice formula presented in the first section of this Chapter.

Why is it wise to keep customers' due dates as late as possible? Because it provides slack for those customers, who really need an earlier due date than the one indicated by the method given in figure 19.1. However, this approach requires time-phased representation of the capacity requirements of the current backlog of customer orders. It becomes necessary to keep track of empty slots of capacity, which can still be filled with orders. This picture of slots with capacity available to promise for new orders requires a changeover plan for important machinery.

Consider, for example, the fine-paper manufacturer of Chapter 11. In this case, the paper machine is a highly expensive piece of machinery with sequence-dependent set-up times. This company has decided to determine its production runs in advance, in order to be able to obtain an efficient changeover schedule. Also, the total run-length has been fixed in order to control the fraction of changeover time in relation to the total productive time of the paper machine. Of course, the precise amount of, say, 120 grammage paper is not completely fixed in advance. In other words, if the run is still largely empty, but 120 grammage paper happens to be fully booked, it is always possible to borrow some capacity from, say, 130 grammage paper. However, if the whole run is booked, a customer order has to move to another run. Also, the length of the 120 grammage run in calendered paper is limited because of the limitations on the *total* amount of calendered paper in each run.

Generally speaking, changeover schedules can be determined at the level of:

- aggregate production planning
- factory coordination
- production unit control.

When changeover sequences are determined at the level of aggregate production planning, this sequence becomes a constraint in quoting due dates for customer orders. Similarly, such a sequence becomes a constraint in planning work orders at factory coordination level. This is clearly illustrated by the 10-day production cycle in the case of Chapter 11.

When changeover sequences are determined at the level of factory coordination, they are usually established in interaction with accepting customer orders and with work-order planning. This means that planning activities become rather complex at this level of decision-making. In the case of the fine-paper-manufacturing company, no changeover schedules are determined at the level of factory coordination. However, incidental rescheduling of work to be calendered interferes with promised due dates, and should therefore be checked!

When changeover schedules are determined as part of production unit control, then the effect of this production unit control on lead times has to be estimated at factory coordination level. In the case of the fine-paper-manufacturing company, changeover schedules in machines such as the sheet-cutter are part of production unit control, and these effects contribute to the estimated lead times of customer orders.

The distinction between factory coordination and production unit control
The distinction between aggregate production planning and factory coordination has been introduced above. The distinction between factory coordination and production unit control will be discussed now.

Factory coordination is concerned with two main issues:

- external: negotiation with customers and suppliers
- internal: planning of work orders within production units and planning "inventories" between production units (the term "inventories" is put in inverted commas because they take the form of backlogs in pure make-to-order business).

Production unit control is concerned with the due dates of work orders, with efficiency of production, with the quality of the products produced and with the quality of working conditions. Production unit control is closely related to the APICS definition of production activity control, with one exception: the concept of *production units* plays a key role here. For this reason we will devote some attention to this concept here.

A production unit is a self-contained part of a factory. Self-containedness means that the production unit is able to accept or refuse work orders and that accepted work orders are realised with an agreed throughput time. Self-containedness also has another meaning: the capacity constraints and lead times should be simple, so that they can be modelled and approximated by factory coordination without the need for much communication.

The concept of production units is needed in order to enable factory coordination to react fast to disruptions and external events. In particular, negotiations with customers and suppliers require that throughput times through the factory or parts of it should be easily calculated.

The material aspect and the capacity aspect of production control
In many factories, it is wise to distinguish more than one production unit. The boundaries of these production units determine the *structure* of the production system and the points where work orders start and finish. This subject is discussed in more detail in section 19.3.

Factory coordination consists of planning work orders through production units in such a way that lead-time norms in these production units are not violated. Planning the same material flow through subsequent production units can be considered as the *material aspect* of production control. In terms of the Walrasian model introduced in Chapter 5, it constitutes vertical coordination. *Between* two subsequent production units, "inventory" will occur. This inventory is necessary because the sequence of work orders at the dominant machine in the first production unit may differ from the sequence in the second production unit. As mentioned earlier, this "inventory" in customer driven production is a set of semi-finished materials for the backlog of customer orders which have been accepted but not yet shipped.

The "sequences of work orders (or sequences of set-ups) at a machine" can be considered as the *capacity aspect* of production control. In terms of the Walrasian model, introduced in Chapter 5, it constitutes horizontal coordination. The "inventory" between subsequent production units is mainly caused because the capacity aspect interferes with the material aspect.

Production unit control consists of prioritising operations and allocating capacities on the shop floor in such a way that production proceeds within the norms of quality, efficiency, lead times and working conditions. The distinction between the material aspect and the capacity aspect can also be made at the level of production unit control. The material aspect is represented by the sequence of operations to be performed on the same work order, whereas the capacity aspect is represented by the wish to produce efficiently. Queues on the shop floor mainly occur because the material aspect interferes with the capacity aspect.

Designing the production process structure 249

The fact that the material aspect and the capacity aspect both occur at two levels of control raises the issue of how the different production units of a factory can be distinguished. In other words, why should a particular work order in the factory stop somewhere and the materials put into inventory and wait until another work order is released before material processing can move on? This question is answered in the next section.

19.3 DESIGNING THE PRODUCTION PROCESS STRUCTURE

We are now in a position to outline the reasons for distinguishing several production units in a factory. This outline will be given for customer-driven process-oriented factories, but it can be generalised to other manufacturing systems.

Self-containedness

The first consideration in distinguishing production units is their self-containedness. It should be clear which resources belong to the PU, and humans within the PU should preferably have skills for multiple tasks within the PU in order to enable flexible reaction patterns to absenteeism or quality problems. The lead times of products within the PU and its capacity should be easily understandable and computable.

These requirements would perhaps lead to the suggestion that each machine should be another PU. In most factories, however, this would result in a very complex materials coordination problem. PUs should therefore be chosen as large as possible, taking into consideration the other arguments given below.

The customer order decoupling point

The second reason is related to the customer order decoupling point (CODP). This point was introduced in Chapter 6 as the point from which the flow of material is driven by customer orders. Normally there is an inventory located at the CODP. Consequently, it is natural to expect that the CODP separates two production units: a forecast-driven PU and a customer-order-driven PU.

One point should, however, be made. In Chapter 6 it was to some extent suggested (as in other texts) that the CODP is a unique point for a whole factory. This idea is too simple. In customer-driven workflow-oriented companies particularly, the CODP is often dependent on the type of product, the customer, the amount ordered, the price agreed or some other variable. This means that at several points in the material flow where the CODP could be located, a separation between two production units is desirable.

Material considerations
In many factories, materials converge and diverge. In a converging situation, such as in blending or assembly, it is risky to allow for a large variance in lead times through one PU. For example, in pharmaceutical products it may take months to produce the key substance through patented processing, whereas packaging materials can be obtained in a few days. In such a situation it is seldom wise to have processing and packaging in one PU, as the material problems are highly uncertain at the time when processing starts. Market conditions and production yield are better known when processing is ready and the key substance is available. Therefore, it is easier to split the factory into a processing PU and an assembly PU. This argument also holds in customer driven converging production systems, such as printing industries if there is reduction in uncertainty during the long lead time of the main item.

In a diverging situation, the material available can be used potentially for many applications. This is a reason for having a stocking point just before the divergence, in order to keep as much freedom as possible at material coordination level for material usage. In customer driven flow shops, this argument often holds upstream of the CODP. It seldom holds downstream of the CODP, because customer driven materials are seldom suitable for many purposes.

Capacity considerations
Capacities are often more important than materials in customer driven flow shops because the capacity structure is more complex than the material structure. Capacity structure relates to changeover sequences and other constraints at the machines, to the flexibility of humans, and to the utilisation rate of the resources. Bottleneck resources can best be located just after the boundary of a PU, in such a way that work order release to the bottleneck is done by factory coordination. Lead times in the PU after the bottleneck should not be a problem. This solution is clearly possible for the paper machine and first steps beyond the PM in the case of Chapter 11.

Machines with dominant changeover sequences are in many respects similar to bottlenecks. These machines should sometimes be planned at aggregate production planning level or at factory coordination level. It is difficult to have such machines in one PU, because a single order release policy cannot feed two subsequent entirely different machines without a substantial increase in stocks.

Finally, resources in one PU should be changeable within the same time frame. Machines or operators which cannot be placed in a three-shift routine should not be placed in one PU with other resources that are more flexible.

These rules give a brief indication of the rules distinguishing production units. We shall now explain how these rules apply in the case of Chapter 11.

19.4 PRODUCTION CONTROL IN MORE DETAIL: THE FINE-PAPER FACTORY - PROPOSALS FOR IMPROVEMENT

In Chapter 11 we discussed the activities of a paper factory and the problems that make a smooth flow of goods in the factory difficult. Before discussing proposals for improvement, it is appropriate to discuss the Grai grid for this case. In figure 19.2 the goods flow control structure of the case study is summarised in terms of a Grai grid. The various levels of decision-making (annual planning, aggregate production planning, PM planning) are clearly indicated by the rows of the Grid. The different aspects of decision-making are also indicated. Annual planning is concerned with the acquisition of resources and with contracts for raw materials. Aggregate production planning employs available resources in order to establish the (cyclic) aggregate production plan. Customer orders are accepted against the aggregate production plan and they become the input for PM planning. PM planning involves a final resource check and ordering materials. It results in combinations to be released to the PM, it implies plans for the coater and the calenders, and through the trimming program results in letters to be produced by the slitter-rewinders.

Figures 19.3 a, b, and c depict Grai grids for shop floor control at the paper machine, the coater and calenders, and in the finishing department. At the paper machine the PM plan is, in principle, followed, but the actual production situation is taken into account - as well as requests for changing the sequence slightly in order to solve downstream capacity problems. The situation at the calenders is

252 Production control in workflow-oriented make-to-order firms

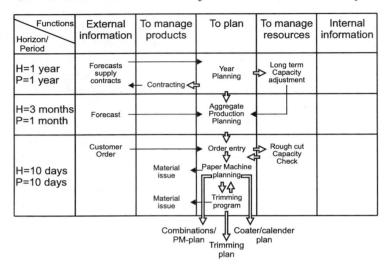

Fig. 19.2 Grai grid for aggregate production planning and factory coordination

identical, except for the fact that a resource allocation decision has to be taken. In the finishing department, an additional decision is first included, viz. the decision on collecting reels which have to be processed together into sets.

Although the paper machine should be the bottleneck machine, unintentional (sometimes temporary) bottlenecks occur elsewhere. The machine group that causes the greatest trouble is sheet cutting.

In order to solve the problems at the sheet-cutting machines, the production management decided to expand sheet-cutting capacity. This will result in an expected overcapacity at the sheet-cutting machines of approximately 4%. This overcapacity will mitigate some of the problems described in Chapter 11. For instance there will probably be less undesired transport of paper from one factory to another. However, an overcapacity of only 4% will not result in great flexibility; space requirement problems and fluctuations in supply remain. In many respects the sheet-cutting machines remain bottlenecks and therefore should be treated as bottlenecks.

A possible solution is to split up the factory into two production units and create a decoupling point between the slitter-rewinders and the sheet-cutting machines. The reason for this suggestion lies in the fact that the maximum throughput for the slitter rewinders occurs for high basis weights, while the maximum throughput for the sheet cutters occurs for the low basis weights. Now, remember that the entire factory

The fine-paper factory - Proposals for improvement 253

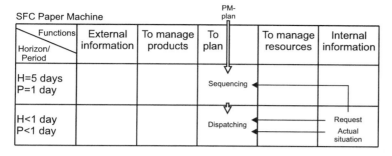

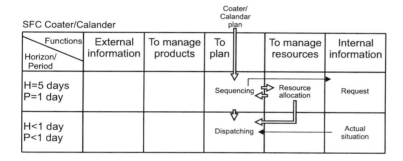

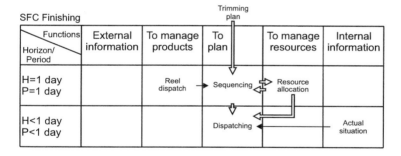

Fig. 19.3 Grai grids for paper machine, coater/calenders, and finishing

changes from high to low basis weights and back every 10 days. Obviously, the factory cannot avoid having a large queue for the sheet cutters at the end of the high-basis-weight period — and it cannot avoid empty queues at the end of a low-basis-weight period.

What sense does it make, then, to split the factory into two production units? After all, the lack of space before the sheet cutters remains. There are many reasons:

Firstly, if there is a *planned* queue in high basis weights before the sheet cutters, the reels that should be moved to higher floors could also be planned in order to keep space for new orders. If it were the task of Factory Coordination (rather than Shop Floor Control) to decide on the orders to be moved upwards, this task could be related to the time-slack of customer orders. The planned queue becomes a managed backlog. Such a procedure could avoid the situation where orders become urgent because they have been forgotten.

Secondly, a new decoupling point could enable the creation of a shift in the customer order decoupling point (CODP) for a limited set of orders. After all, some of the reels with paper in each combination of basis weight and quality are far more common products than others. About 15% of the reels in each are pretty standard, in the sense that a reel of this width is produced in almost each production cycle. If these reels were kept on stock, the company could perhaps solve several problems at once:

- In periods where the sheet cutters have a lack of work, some of the customer orders could be produced before the paper has arrived from the slitter-rewinders simply by taking the appropriate reels from stock. The stock can be brought back to its original level by moving reels produced by the slitter-rewinder back into stock later.
- In periods where the slitter-rewinders produce at full speed, there is a possibility for avoiding storage problems before the sheet cutters by giving priority to reels which have to be moved upwards anyway.
- In the case of quality problems (paper damage) before the slitter-rewinder, it may be possible to change the cutting programme at the slitter-rewinder in such a way that reels produced for stock are taken out of the combination to be cut, leaving room for reels with non-standard widths. Of course, the reels which are not produced in the current production cycle are added to the volume of the next.
- In the case of quality problems (paper damage) after the sheet cutters, the work can be redone by taking another reel with the same quality and basis weight from stock and cutting the appropriate sheets from this reel. However, in this case the width will normally have to be reduced through guillotine cutting, which produces extra scrap. This possibility therefore requires a careful balancing of objectives.
- A stock point after the slitter-rewinder would enable Sales to promise a substantial reduction in lead time, incidentally.

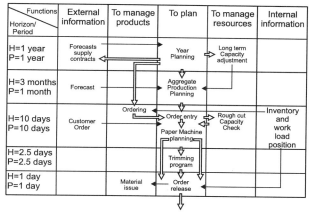

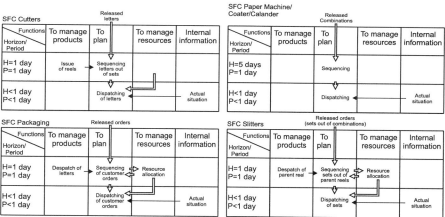

Fig. 19.4 Grai grid at goods flow control level for the fine-paper factory after introduction of an intermediate stock point

Thirdly, a split into two production units makes it easier to release work orders to the factory. In the old situation there were no real work orders. Rather, the factory was given a set of schedules for different machines. With two PUs, the first PU receives work orders to produce reels, mostly for customer orders but also for stock. The second PU receives work orders to produce customer orders from specific reels, preferably packaged.

It could be argued that the introduction of a sheet-cutting order release would take away the autonomy of shop floor workers. This is a misconception. After all, releasing work orders to the second production unit still leaves room for setting the details of allocation and the collection of sets. However, the factory can use many opportunities elsewhere to improve the situation. Stimulating multi-task capabilities is one of them. Equally important is the attitude towards quality control. If the workers in the shifts were to have the authority to solve quality problems by themselves, a considerable source of disruption in planning would be removed.

The introduction of the stock point after the slitter rewinder changes the decision structure at goods flow control level considerably. This is illustrated in figure 19.4, which should be compared with figure 19.2.

20

Production control in product-oriented assemble-to-order manufacturing

20.1 INTRODUCTION

In this chapter we will proceed with one of the best-known types of customer-driven manufacturing, viz. assemble to order. We will discuss in detail the concepts required for Medicom, following the case description of Chapter 10. The problems of Medicom are partly due to the classical way of using bills of material and other product descriptions. We have argued in Chapter 16, the *generic product modelling concept* may solve this part of the problem. However, an improved product-modelling concept should go hand-in-hand with a better production control concept. This concept is to be discussed now. Implementation of these concepts through information technology will be discussed in Part D, specifically in Chapters 26 and 27.

In the previous chapter, Chapter 19, we introduced a number of issues which should be resolved in designing production control systems. These issues are:

- the role of Aggregate Production Planning
- the interface between Production and Sales
- the distinction between Goods Flow Control and Production Unit Control
- the coordination of the material aspect and the capacity aspect.

Furthermore, Chapter 19 paid attention to the design of the *production process structure*, which refers to the definition of production units and the selection of GFC-controlled items. A most important design choice is concerned with the customer order decoupling point (CODP).

All these issues play a role in this chapter as well as in subsequent chapters of Part C. However, these issues will not be discussed each at the same level of detail here. We will largely neglect in this chapter the choice of production units, and assume in Section 20.2 that Medicom has decided to use the *subsystem* level as the customer order decoupling point. It should be recalled, that subsystems at Medicom are complex assemblies such as a user console. This choice of the CODP will lead to a relatively straightforward production control concept.

Next, in Section 20.3 we will investigate the consequences of moving the CODP more upstream, so that manufacturing of subsystems from components will become assemble-to-order manufacturing in itself. This will lead to some additional problems in customer-driven production control, for which the generic bill-of-material concept of Chapter 16 provides a solution. This solution is discussed for the case of Medicom, as a preparation for the discussion of IT concepts for Chapter 26.

Section 20.4 discusses briefly how the concepts of Section 20.3 could be realized in a more repetitive customer-driven manufacturing environment, such as Automotive. Section 20.5 presents conclusions.

20.2 CUSTOMER DRIVEN ASSEMBLY OF SYSTEMS FROM SUBSYSTEMS AT MEDICOM

Let us assume that Medicom has decided to position the CODP at the level of subsystems. This means in effect, that most factories of Medicom apply a make-to-stock manufacturing concept, such as MRP II (cf. Chapter 27), and that there are a few customer-driven factories where systems are assembled. Similarly, the Product Groups where subsystems/components are developed and marketed will keep the classical concept of standard products manufacturing. Only the Systems Product Groups will develop and market concepts for customer-driven integration.

This approach will lead to a relatively simple and transparent solution for assemble to order, that is illustrated by the Grai-grid of figure 20.1 (cf. Chapters 18 and 19):

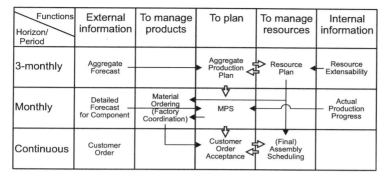

Fig. 20.1 Grai grid for customer-driven assembly of systems at Medicom

A key role is played in this concept by the Master Production Schedule (MPS) for manufacturing of subsystems. This MPS is established in a monthly decision cycle, and it is based on several inputs. First of all, detailed forecasts per subsystem are required. Second, the Aggregate Production Plan serves as a constraint in establishing the MPS. Third, the actual progress of the production and material supply acts as an input for the MPS. However, the only formal way of taking this information into account is by consideration of the previous MPS.

An important requirement for the MPS is that it is *balanced*. The term "balanced MPS" refers to a situation where the MPS for different items at the CODP is matched with the expected demand. This matching refers to two points:

- *Balance of mix*: different items which have a similar function should be produced according to the probability distribution of the family demand. For example, if there are two types of user consoles, and these are sold in a distribution of 25% vs. 75% then this fact should be reflected in the MPS.
- *Balance of volume*: If different families of components/subsystems are used together to be assembled into a system, then these families should have identical consolidated production figures in the MPS. For example, if one user console is produced per day, then also one generator should be produced per day.

These requirements are met relatively easily in the case where subsystems constitute the CODP. The first requirement - balance of mix - is met by a so-called "percentage bill of material" which relates alternative subsystems from one family to the family plan.

The second requirement - balance of volume - is met, if all families of subsystems are planned according to the Aggregate Production Plan. Both requirements can be met by a family bill-of-material structure as shown in figure 20.2. Such structures are part of the generic bill-of-material concept introduced in Chapter 16.

Aggregate Production Planning at Medicom is done at systems level, for example with respect to a family Cardiovascular Systems. This plan is exploded to subsystem families by means of *family bills of material* (cf. figure 20.2). The aggregate plan for subsystem families is translated into a production demand for individual stocked types of subsystems through *percentage bills of material* (cf. figure 20.2).

The explosions over the BOM structures of figure 20.2 can be enhanced by added functionality of decision support software. With respect to the family BOM, off-setting the aggregate plan over the assembly leadtime of entire systems is required. Furthermore, some type of manual netting of the (aggregate) demand at subsystem family level against (aggregate) inventory may be necessary.

With respect to the percentage BOM, the aggregate plan at subsystem family level has to be translated into a real MPS of individual subsystem items. The same holds true for forecasting figures.

On the longer term, the sales forecast and the MPS of subsystems can be derived straightforwardly from the aggregate plan at family level. On the short term, however, the MPS-planner should react manually

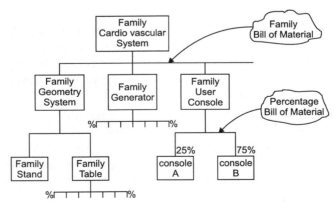

Fig. 20.2 Aggregate Production Planning with family BoM and percentage BOM

to variations in actual demand. Therefore, the nearby MPS may certainly deviate from the percentage distribution of the percentage BoM.

Similarly, forecasts should react to the actual demand, preferably in an automated way. Forecasts for individual subsystems should also satisfy the constraints stated in the aggregate plans. Therefore, this forecast is sometimes called a "production forecast" in MRP II. One way to satisfy the aggregate-planning constraints is to derive forecasts for individual subsystems from the "available-to-promise" information at family level, using percentage bills. After all, the "available-to-promise" information at subsystems family level satisfies the aggregate planning constraints at subsystems level.

We are now in a postion to explain figure 20.1 in more detail. The 3-monthly Aggregate Production Plan (APP) is based on an aggregate forecast. The APP-decision is taken together with a Resource Plan, for which the extensibility of the current resources should be known. The Aggregate Production Plan should not only plan resources in the subsystems/components factories, but also in assembly and installation, despite the fact that these phases represent the customer-driven part of the supply chain. Furthermore, the Aggregate Production Plan specifies the volume of production of a particular family of systems. This leads to constraints on the MPS of subsystems. For example, if the APP specifies one system per day, then the consolidated number of user consoles (belonging to different types) as planned in the MPS should also equal one per day.

The MPS for the subsystems drives MRP-calculations. These should be confronted with detailed capacity checks at the level of factory coordination in the factories where subsystems/components are manufactured.

The MPS is also the basis for an Available-to-Promise (ATP) calculation at subsystem level. The ATP is the basis for Customer-order Acceptance (COA). Customers can select systems by choosing parameters on choice sheets (see Fig. 16.3).

Customer orders can be entered on a continuous basis. Promising due dates for these customer orders should happen in direct interaction with final assembly scheduling and installation planning. This requires insight in the available capacity and the load of the production units which are supposed to deliver final assembly and installation activities.

The Grai grid of figure 20.1 represents a decision structure. Now consider figure 20.3, which shows this same decision structure superimposed on the goodsflow. For reasons of simplicity, the planning of capacity resources has been omitted here. However, figure 20.3 shows more clearly what the difference is between goodsflow control upstream and downstream of the customer order decoupling point (CODP).

Upstream of the CODP, the work is released to production units by MRP work orders. These work orders act as scheduled receipts for standard items (cf. Chapter 27), using standard bills of material and standard routings. Downstream of of the CODP there are no standard items. Work is released to the production units as customer-specific work orders with customer-specific items and customer-specific routings. The generic bill-of-material concept introduced in Chapter 16 provides an excellent way to create these customer-specific work orders. A similar concept can be applied to routings.

However, merely printing these work orders with bills of material and routings is not sufficient for Medicom. After the creation of these work orders they have to be maintained. The work orders are needed for costing purposes, for quality control, and as reference documents for later service.

Costing requires that a specific anticipated cost price can be established for each system, which can be compared with actual costs incurred. Quality control requires at Medicom that each system is documented individually, including test reports for subsystems and components. This requires also lot traceability.

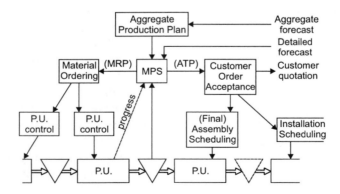

Fig. 20.3 Decision structure superimposed on the goodsflow

20.3 CUSTOMER DRIVEN ASSEMBLY OF SYSTEMS AND SUBSYSTEMS FROM COMPONENTS AT MEDICOM

In this section, we assume that the CODP for Medicom is no longer situated at the level of subsystems, but at the level of components. Recall that subsystems are major assemblies such as a generator, an image intensifier, or a user console. The reason for shifting the CODP upstream lies in the fact that these subsystems are becoming variants of families in themselves. The increasing variety makes it unattractive to keep the different subsystems on stock.

In this case, it becomes increasingly difficult to maintain a balanced MPS. Recall that a "balanced" MPS means both balance of volume and balance of mix. The balance of volume is obtained through family bills, whereas the balance of mix is obtained through percentage bills. When the CODP is shifted upstream, especially the balance of mix becomes much more cumbersome. The following problems emerge:

- *Dependency*
  The percentage bills may be interpreted as probabilities of choices by customers. However, the percentages used in different parts of a BOM structure such as given in figure 20.2 are subject to constraints. This means that percentages in different percentage bills cannot be chosen independently.
  N.B. This fact was also present in the case discussed above, but perhaps it could be neglected there: it could be argued in the previous case that a percentage bill for *all* allowed combinations of subsystems would still yield a workable solution.
- *Redundancy*
  The percentage bills may easily contain redundant information. The distribution of values of the parameter voltage over 220 vs. 110 Volt will occur probably in each subsystem. But there are many similar choices which have to be maintained, probably in different subsystems. These percentages have to be updated based on realized sales figures in a consistent way.
- *Netting*
  Although the CODP is *in principle* moved upstream to components, it may occur that subsystems are produced on stock. The MPS may have to be netted against available inventory for such subsystems.
- *Constraints*
  The constraints have to be represented in a way which avoids

violation during configuration. In other words, the constraints should be checked during the customer-order acceptance procedure. Furthermore, this should be done in a way which is consistent with the first point made above. Finally, constrains should be specified only once in order to keep the systems transparent and maintainable.

- *Available to promise (ATP)*
  Customer orders are accepted and delivery dates are quoted after checking the available-to-promise figures of MPS items. As long as the MPS items represent subsystems, this way of working is reasonable. However, when MPS items are components, ATP figures have to be checked for thousands of items for each customer order. Clearly, more robustness and speed would be obtained if ATP numbers could be available for (subsets of) families at subsystems level.

The solution for all these problems adds to the requirements for a robust generic bill-of-material system. These requirements are, roughly speaking, that families of subsystems may be components of several families of systems and that families of components may be part of several families of subsystems. Furthermore, it should be possible to configure each subsystem by itself. More precisely, the generic bill-of-material system should support:

- the definition of product families at all levels of the hierarchical product structure
- the specification of properties for each product family (at any level in the BOM)
- the translation of a set of product properties at system-family level into a set of product properties at subsystem- or component-family level
- a specification of the combinations of product properties which are valid (constraints)
- the specification of planning percentages for each property of a family or for a limited set of combinations of properties (as determined by the constraints)
- the definition of individual products within a product family.

Such a generic bill-of-material concept has been developed by Hegge [1995]. In this concept, percentages can be maintained in such a way that configuration constraints are taken into account and that the

percentages do not have to be specified redundantly. The des
of the GBOM in Chapter 16 is consistent with these requirements. More
details about this solution are given in Chapter 26.

20.4 CUSTOMER-DRIVEN ASSEMBLY IN REPETITIVE MANUFACTURING

The case of Medicom is a typical product-oriented assemble-to-order case. However, it is by nature an *industrial project* type of business. Modern automotive assembling industries would be characterized also as product-oriented assemble-to-order manufacturing, be it of a *repetitive* nature.

There is one important distinction between industrial projects and repetitive manufacturing. As we have seen, industrial projects typically require that each customer order is documented as a *project* in terms of costing, quality control and as-built configuration.

In repetitive manufacturing, such a requirement would cause too much burden. From a costing point of view, there is an interest in cost centers but not in the costs of individual products. From a quality point of view, there is an interest in the traceability of certain lots of components in relation to serial numbers on the assmbly line. However, there is no requirement to keep the as-built documentation of each individual vehicle.

Therefore, the leading principle in automotive should be, to generate documents whenever they are needed, but not to store and maintain them in relation to the individual vehicle.

For the purpose of later reference and reuse of the generated documents, it is necessary to have an excellent version and status control. After all, it should be possible after dozens of years to reproduce any document produced earlier. This means that engineering data management should not only be concerned with classical bills of materials and drawings, but especialy with the generic data and constraints, introduced in Chapter 16. This point returns in Chapter 28.

20.5 CONCLUSION

This chapter has shown a production control system for Medicom. It has been shown, that product-oriented assemble-to-order manufacturing requires a hybrid concept of customer-driven manufacturing and

MRP II. The MPS will be derived from aggregate plans through family- and percentage-bills of material. These bill-of-material concepts can be implemented in a generic bill-of-material system, such as introduced in Chapter 16. This system will be discussed in more detail in Chapter 26.

The Aggregate Plans are based on forecasts and other considerations, such as marketing management policy, earlier plans, and available inventories or backlogs. The customer orders are accepted by checking the available-to-promise figures, which are based on the MPS and customer orders.

Production control downstream of the CODP is mainly based on customer orders. However, production capacity may restrict the planned volume of production by using a kind of available-to-promise logic for capacity. In cases such as Medicom, production documents for each customer order will be generated and stored separately, as individual projects. In more repetitive production environments, such as automotive assemblers, the production documents will be generated on an as needed basis, but not be stored and processed individually.

REFERENCES

1. Erens, F.J., *The synthesis of variety*, Ph.D. Thesis, Eindhoven University of Technology, 1996.
2. Hegge, H.M.H.
   Intelligent Product Family Descriptions for Business applications. Ph.D. Thesis, Eindhoven University of Technology, Eindhoven, The Netherlands, 1995.
3. Van Veen, E.A., *Modelling Product Structures by generic Bills-of-Material*, Elsevier, 1992.

21

Production control in engineer-to-order firms

21.1 INTRODUCTION

We will start this chapter with the description of some production control design principles that can be applied to our engineer-to-order case situation (Chapter 9). Using these design principles, a production control framework will be developed and described in the subsequent sections. The different production phases and production units in the production process are defined as the first step (Section 21.3). In connection with this, a distinction is made between Production Unit Control and Goods Flow Control. The various production control functions are then developed as the second step (Section 21.4). The third and final step is to develop the decision structure within this control framework (Section 21.5).

21.2 PRODUCTION CONTROL DESIGN PRINCIPLES

Bertrand et al. (1990) have developed four general design principles for designing a production control framework for production environments. These design principles are:

- the decision structure should be seen as the basis for a good control framework (this point has been discussed in Chapter 18);
- a distinction should be made between goods flow control and production unit control;
- a distinction should be made between the detailed item-oriented control and the aggregate capacity-oriented control;

- special attention needs to be paid to the interface between Production and Sales.

The last three points will be discussed in more detail.

Goods flow control versus production unit control

Two hierarchical production control levels can be identified: the production unit control (PUC) level and the goods flow control (GFC) level. Goods flow control is concerned with the overall coordination for a chain of *production phases*. Each production phase then becomes a set of operations with an input material or product which is transformed to an output material or product. Subsequently, the production phases are assigned to a so-called production unit (PU). A PU is an organizational grouping of resource capacities (see for characteristics: Chapter 14).

The use of GFC provides only for the coordination of the production phases in the primary process. At this production control level the total primary process is defined only in terms of a set of production phases with relationships between these phases (see also Chapter 14). Coordination of the production phases at the GFC level is accomplished by releasing work orders to the PUs for carrying out specific production phases to produce GFC items. A work order is therefore an instruction to initiate and complete a specific production phase. Each individual PU is responsible for completing all of the work orders assigned to it in accordance with the established agreements. It should be obvious that the goods flow control cannot release work orders for processing by a PU without taking into account various factors such as the availability of resource capacity and materials. This release decision is based on aggregated data at a higher level, however. An important design aspect in this connection is the identification of the different PUs. The autonomy of the individual PUs and the importance of PUC both increase when there are fewer PUs in the transformation process.

Aggregate Production Planning versus Operational Production Planning

The following two production control aspects can be identified with respect to the coordination of PUs at the goods flow control level, each with its own production control horizon:

- the coordination and matching at an aggregate level. This involves matching the available resource capacity to the capacity require-

ments. This is a medium/long term coordination activity primarily based upon aggregate data. We will refer to this production control aspect as *Aggregate Production Planning (APP)*;
- the coordination and matching at the detail level. This involves the timing of work orders or, in other words, the periodic assignment of available resource capacity to individual products. We will refer to this production control aspect as *Operational Production Planning (OPP)*.

Interface between Production and Sales
The objective of goods flow control could be formulated as coordinating the interface between Sales and Production, taking the resource capacity and the timeliness of product delivery into account. Sales is responsible for ensuring sufficient product demand and for accepting customer orders. Production is responsible for providing production capacity and ensuring that the customer order is completed on schedule. At certain points in time, however, both Production and Sales will be confronted with limitations and requests which conflict with each other. These situations occur periodically when there is an imbalance between the required and the available resource capacity due to a certain amount of inflexibility in the production capacity, the current workload levels and the available stock. For this reason it is important that the interface between Production and Sales is included within the scope of the control framework.

21.3 STRUCTURING THE TRANSFORMATION PROCESS

A general description of the primary process typically found in an engineer-to-order plant was presented in Chapter 9. We will now describe how this primary process can be structured from a production perspective based upon a number of design principles.

Three design principles can be identified for structuring the primary process for the purpose of production planning and control (refer also to Chapter 14). This primary process involves a specific set of production steps which are needed to manufacture a finished product. Using these principles, a number of steps can be identified to restructure the primary process. These steps are:

- defining the operations;

- identifying the GFC items and production phases;
- establishing the production units.

The relationships between the definition of operations and the primary processes are discussed in more detail by Bertrand et al. (1990). Here we will concentrate more on the identification of GFC items/production phases and the establishment of production units based upon the case situation.

Identifying the GFC items and production phases

The production control issues can be divided into GFC issues and PUC issues through the creation of production phases and GFC items. Six production phases and three production units were defined for the engineering process in Chapter 14. We shall therefore focus on the physical production process. Two extra production phases and related GFC items can be identified by applying the previously-mentioned criteria.

The first production phase is concerned with the manufacturing of the components and basic assemblies (= the GFC items for this phase) which are required for the final assembly of the product in the next phase. Activities which precede the physical manufacturing stage such as preparing the production documentation (process planning) for the various components and ordering the necessary materials and tooling are also included in this production phase. Separate work orders are issued for the manufacture of the various components due to the (assembly) structure of the product. Not all of the components will be required at the same time for the assembly activity. Therefore, it will be necessary to issue separate work orders for the manufacture of individual components to ensure that certain components will not arrive too early at the assembly location. It will be desirable to report updated information at the GFC level at this point to allow for a proper coordination of the component manufacturing and assembly activities.

The second production phase is the assembly of the finished product (= the GFC item for this phase) from the various components. The assembly planning and the preparation of assembly instructions, of course, precede the physical assembly process.

Establishing the production units

Six production units are needed based upon the production phases defined above and the criteria for establishing such production units.

In Chapter 14 we already defined the Product Engineering PU and the Detail Design PU. The production phases of "manufacturing components" and "assembly of the finished product" are each assigned to two separate production units. The manufacturing of components in the case is distributed over two PUs, namely, a Milling PU and a Stamping PU. The assembly activities are divided into two PUs, namely, Pre-assembly and Final Assembly.

The Process Planning PU indicated in Chapter 14 does not appear as part of the total transformation process. This activity is carried out as part of the manufacturing and assembly of components in the case (see Chapter 22).

The concept of a sub order is of primary importance in connection with the delegation of these control responsibilities to the PUs. A manufacturing sub order, for example, consists of a set of components to be produced. Sub orders are assigned to PUs at the GFC level and arrangements are made with the PUC concerning the internal throughput time for these sub orders. The delivery lead time for ordering materials and the throughput time for carrying out the process planning are taken into account. The sub order represents the lowest level of aggregation for the GFC in connection with controlling a customer order. Sub orders are scheduled based upon the aggregate information. For this reason, the GFC is not interested in detailed information such as process planning data and material requisitions concerning the components within the sub orders. This responsibility belongs to the PUs. In this way the GFC only needs to be concerned about the situation at the sub order level.

21.4 DEVELOPING THE PRODUCTION CONTROL FRAMEWORK

At this point it is useful to investigate the characteristics of GFC in more detail. Three production control aspects can be identified within the GFC (refer also to Section 21.2):

- Aggregate Production Planning (APP);
- Operational Production Planning (OPP);
- the interface between Production and Sales.

Aggregate Production Planning (APP)

An Aggregate Production Planning (APP) activity is included within GFC which is independent of the operational control of customer orders. This concerns the medium-term matching of required resource capacity with the available capacity. The available capacity is adjusted as much as possible to be able to meet the estimated future capacity requirements through the use of, for example:

- arrangements with external suppliers for outsourcing capacity;
- fewer or extra work shifts;
- temporary employees to provide additional assembly capacity, etc.

Operational Production Planning (OPP)

In addition to APP, the operational coordination of the production phases of the customer orders is covered by GFC. This production control aspect is also referred to as Operational Production Planning (OPP). This involves the coordination of materials as well as the capacity scheduling.

The characteristics of OPP are different in the different parts of the ttansformation process. Specifically, production control of the customer orders at the start of the process (i.e., during the non-physical processing stage) is carried out based upon aggregate data. This is due to the relatively large amount of uncertainty at this point and the lack of detailed information about the product to be manufactured. The OPP function focuses primarily on the resource capacity aspect at this stage. The material aspect here is limited to the acquisition of the critical materials and components with a long delivery lead time.

As more product information becomes available during the non-physical stage, the global network plan is modified and details are added. This global network plan is also used to monitor the progress of each individual customer order. The production control activities are similarly performed at a more detailed level and the material aspect receives more attention within the scope of the OPP activities when additional product information becomes available. This essentially concerns the coordination of materials and the determination of relative priorities for the work orders in the PUs during the physical processing stage. All of the product details are known by the time that the process planning has been completed for a customer order. A detailed plan is prepared for the various work orders associated with a given customer order, based upon the global network plan with the aggregated activities. The work orders are then released in collaboration with the heads of the various PUs.

Even after the detailed information about a customer order is known, it is still convenient to track the progress of a customer order during the physical processing stage using the customer order network. The aggregated data used in this network provides a good basis for tracking the progress of each customer order.

The interface between Production and Sales

In view of the fact that the customer orders arrive at the start of the transformation process and that the degree of uncertainty is the highest at this stage, the most important production control decisions are taken at the GFC level. The operational arrangements between Sales and Production, in particular, are important at this stage and are included as an explicit part of GFC. An open and intensive exchange of communications between Sales and Production is especially important during the tender phase. A global product design specification is prepared based upon specific customer requirements. This is then used as the basis for calculating the price and delivery schedule for inclusion in the quotation. The capacity requirements at each work station are still quite indefinite at this stage, however. Conditions and circumstances may change during the customer order negotiation phase, for example:

- the customer may change his original specifications, with significant implications for the required resource capacities;
- a number of other customer orders may have been accepted while the negotiations were taking place, reducing the availability of future resource capacity;
- the capacity requirements of previously accepted orders appear to be greater than originally anticipated. This reduces the amount of capacity which is available for new orders;
- the order negotiations extend over an extremely long period of time, leading to changes in the original estimates of the total capacity requirements.

Each of these situations can have an effect on the delivery date to be specified in a tender. A continuous revision and reconfirmation of the relevant conditions between Sales and Production is therefore essential. The point in time at which the customer order is actually placed is also an important factor in coordinating efforts in this area. Due to the competitive nature of the engineer-to-order manufacturing business, any tender for new work will generally have a relatively low probability of being accepted. A potential customer normally asks for several tenders from different suppliers in view of the large size of the investment and the type of product (typically industrial machinery). This means that a potential customer may not decide to place an actual order. If he does, however, it is usually not clear when the order will be placed. A long period of time may elapse between submitting the tender and placement of the corresponding order. Nevertheless, by keeping in touch with the potential customer, Sales is often able to provide a reliable estimate of when a tender is likely to result in the placement of an order. Production needs to be aware of this so that orders do not arrive unexpectedly with commitments for delivery which can no longer be realized.

21.5 DEVELOPING THE DECISION STRUCTURE

Up to this point we have been able to identify the major functions with respect to production control. A distinction has been made between an integral Goods Flow Control (global planning) and local planning and control at the PU level. More substance can be given to the various production control functions by developing a decision structure. The decision structure which is applicable to the case situation is comprised of the following four key production control decisions at the GFC and PUC levels:

1. Customer order acceptance and due date assignment (GFC); is a timely completion of the production of the customer order possible?
2. Sub order assignment and PU outsourcing (GFC); which production unit will be manufacturing which components and what part of the work will need to be contracted out (outsourced)?
3. Work order release (PUC); when will the work be released to the production unit?
4. Work sequencing (PUC); in which sequence will the work be performed within the production unit?

These four key production control decisions are discussed in more detail in the following subsections.

1. Customer order acceptance and due date assignment

Control over the throughput times within the individual production units is a necessary but not a sufficient condition for being able to control the throughput time of the whole customer order. This is because the throughput times within a PU are only controlled at the PUC level. The integral coordination between Sales and Production should take place at the GFC level (refer to the previous section). This includes deciding how much effort is to be spent preparing a tender in each specific case, what delivery lead time will be quoted and what the price will be. A good structural and operational interface between Sales and Production is needed.

If we visualize the order acceptance and due date assignment decision as being a regulating faucet, then the quantity of work flowing into the production organization can be controlled by opening and closing this faucet. In practice, a major part of the production and financial control problems within engineer-to-order manufacturing plants could be solved by implementing an effective faucet to regulate the work flow in this way. It is apparent that this key decision deserves more attention than it currently receives in the literature as well as in practice (see also Kingsman et al., 1989). A large number of variables and other factors need to be taken into consideration in making this decision. These factors include the future capacity loading, the relative value of a potential order, the desired delivery lead time, the probability of a tender becoming a firm order and the technical risks. Various business disciplines such as Sales, Engineering and Production are involved in the order acceptance and due date assignment decision. Sales is responsible for determining which price will be quoted in the tender and also plays an important role in compiling all of the customer specifications. Engineering (or Product Development) is responsible for translating the customer specifications into the technical specifications for a manufacturable product. Engineering must also provide an estimate of the technical risks associated with accepting all of the customer's wishes when the order is accepted and the due date is assigned. This is then used by Sales to determine the quotation price and by Production to determine the delivery lead time. This decision is taken at the GFC level since it can be characterized as being an integral decision.

2. Sub order assignment and outsourcing

The first key decision is typically taken when there is still a great deal of uncertainty. Usually only a part of the manufacturing characteristics of a custom-made product will be known at the point in time when such a customer order is accepted and the due date is assigned. Nevertheless, a delivery lead time and a price have already been set. A major part of this product uncertainty disappears when the custom-made product is fully developed within the Product Engineering and Detail Design PUs. A much better estimate of the required quantities and types of resource capacities can be made at this point. Even before the process planning takes place, it is possible to assign the manufacturing of components to a specific PU (assuming that there are two or more PUs for component manufacturing and/or assembly activities) based upon the technical content of the work. The customer order is split into so-called *sub orders* for this purpose. A sub order is defined as being the collection of all of the work associated with a single customer order which is to be processed during a given period of time in a given PU. Each sub order is assigned to a specific PU before the capacity loading is analyzed. If one or more of the PUs have insufficient capacity, then a number of the sub orders could be contracted out to an external supplier.

In many engineer-to-order production situations flexibility with respect to the volume of products in component manufacturing is realized by outsourcing some of the work to suppliers. In view of the availability of volume flexibility in this way, the second key decision has the following two objectives:

- Assigning the sub orders to PUs as quickly as possible to enable an evaluation of the capacity loading situation within the PUs as soon as possible. This evaluation can then be used to determine which sub orders are to be processed internally and which sub orders, if any, are to be contracted out. When sub orders are assigned at an early stage in this way, the PU department heads are then able to evaluate the future demand for their resource capacity and take timely action as appropriate.
- Review of the first key decision in the light of new information which has reduced the level of product uncertainty. Corrective measures will need to be taken in connection with the second key decision if it appears that certain variables have developed in a way which is contrary to the original expectations. Examples of corrective measures are, for example, the use of available slack time, extra outsourcing and the use of internal flexibility.

In view of the fact that delivery lead times for outsourcing tend to be longer than the internal production throughput times, it is extremely important to make arrangements for outsourcing as early as possible. A plant can be more flexible and react quicker to changes in the market when it is able to arrange for outsourcing more quickly than its competitors. Delivery problems typically occur in practice when the work to be contracted out is not released at an early date.

The concept of *sub orders* as used in this Chapter is of primary importance for the control framework presented here because this provides the most important link between the non-physical and physical processing stages in the primary process. The use of sub orders also implicitly provides a vehicle for communicating product information to the component manufacturing unit. All of the component drawings for a given sub order are first completed within the Detail Design PU before the complete sub order is transferred to the PU where the components are to be manufactured. The sequencing of operations within the Detail Design PU occurs at the sub-order level (with priority being given to the sub orders on the critical path in the customer order network). Agreement is achieved between the GFC and the PUC concerning the internal due dates for the sub orders. The PUC of the components manufacturing unit controls its own process planning, materials requisition and manufacturing for the various components included in each sub order.

3. Work order release

The decision structure should be designed to allow for controlling the throughput time of the customer order as much as possible and also for ensuring that the targeted capacity utilization levels can be realized. The throughput time of the customer order is determined by the throughput times of the activities within the various production units which are on the critical path of the customer order. The objective of PUC is to accept work orders and to ensure the realization of the accepted work orders within the agreed throughput time, taking the agreements regarding utilization levels, throughput times and batch sizes into account. A significant amount of research has been done on controlling throughput times within production units; refer, for example, to Bertrand and Wortmann (1981), Bechte (1987) and Wiendahl (1987). These studies have demonstrated that throughput times are dependent upon, among other factors, the workload level found within the production unit. The more work found on the shop floor, the longer the average throughput time will be for newly accepted work orders.

Throughput times can therefore be controlled by controlling the workload level (see for further details: Chapter 23).

The third key decision of releasing work to a PU is the first control decision at the PUC level. This type of decision about releasing work is, of course, applicable to the PUs in the physical processing stages as well as to the PUs in the non-physical processing stages.

4. Work sequencing

An additional (combined) control decision is taken within the production unit which influences the throughput time performance of the PU after the work is released to the production unit. Within each PU there is a certain amount of work present (the workload) which has been completed to various degrees. Each day it is necessary to determine which capacities have to be allocated to which work and in which sequence the work will be completed. In other words, the sequencing of processing the work orders within the production unit must be determined. A multitude of details and conditions are taken into account and anticipated at this stage which could not be included in earlier versions of the plans, such as: combining similar types of work, utilizing alternative equipment and machinery, reallocating human resources from other work stations. A certain amount of flexibility is required on the shop floor to allow for an adequate coordination of

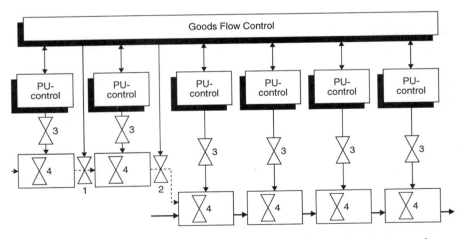

Fig. 21.1 The position of the four key decisions in the production control diagram

the various work orders in view of the uncertainties and the stochastic nature of this type of manufacturing environment.

The fourth key decision, the sequencing decision, is the second decision at the PUC level. The positioning of the four key decisions in the production control diagram at both control levels is illustrated in Fig. 21.1.

The delegation of decision-making responsibilities is explicitly included in the decision structure described here. The first key decision is generally taken collectively by the commercial management and the production management or, occasionally, by the complete management team. The second key decision is taken by the productions manager together with the heads of the production departments. A meeting is held with the heads of the production departments at the earliest possible moment to discuss the implications of the quantity of work which is expected. In this way they are given adequate time to take any measures which may be appropriate. The third key decision is taken by each head of a production department together with the group leaders within his department. Finally, the fourth key decision is taken by each group leader together with the individual workers in his group. As the order progresses along the transformation process, the responsibility for making control decisions becomes increasingly closer to the operations on the shop floor. The requisite authorities for taking these decisions must, of course, also be established.

REFERENCES

1. Bertrand, J.W.M. and J.C. Wortmann, *Production Control and Information Systems for Component Manufacturing Shops*, Elsevier, 1981.
2. Bertrand, J.W.M., J.C. Wortmann and J. Wijngaard, *Production Control, A Structural and Design Oriented Approach*, Elsevier, 1990.
3. Kingsman, B.G., I.P. Tatsiopoulis and L.C. Hendry, A structural methodology for managing manufacturing lead times in make-to-order companies, *European Journal of Operational Research* 40, pp.196-209, 1989.
4. Bechte, W., *Theory and Practice of Load-Oriented Manufacturing Control* 26, pp.375-395, 1988.
5. Wiendahl, H.-P., *Belastungsorientierte Fertigungssteuerung*, Carl Hanser, 1987.

22

Management of tendering and engineering

22.1 INTRODUCTION

From the analysis of the selected case in Chapter 9 it is clear that little or no attention has been paid to the controlling system for customer driven engineering. This may result in situations in which impractical delivery dates are approved and scarce engineering resource capacity is spent on the wrong requests such that a majority of the customer orders are then delivered too late and with insufficient profit margins. The required controlling system is designed in this chapter, based upon the engineering process designed in Chapter 14 and the controlling system of the total transformation process designed in Chapter 21. Based upon this, the design of the controlling system for the customer driven engineering is described in Sections 22.2 and 22.3, respectively.

22.2 THE CONTROL SYSTEM FOR CUSTOMER DRIVEN ENGINEERING

22.2.1 System boundaries for customer driven engineering

The input flows with respect to the production system under consideration are:

- the requests for quotation for a specific machine which are received by Sales from customers. This includes a specification of the customer's requirements;
- customer orders which are received as the result of issued quotations.

The output flows with respect to the production system under consideration are:

- quotations (or rejections of requests for quotation) sent to customers in response to their requests for quotation;
- order confirmations sent to customers, resulting from the receipt of customer orders;
- drawings and bills-of-materials as required for the physical production.

It can be seen from the system boundaries defined above for customer driven engineering in the case situation that the preparation of production documentation (process planning) has not been included within the system. This activity is carried out as part of the manufacturing and assembly of components in the selected case. As a result, no separate PU has been defined for this activity. In the design of the customer driven engineering process in Chapter 14, however, the initial assumption was made that a separate PU would be defined. The decision to eliminate the separate production phase for process planning was based upon the control characteristics of the case situation described in Chapter 9.

In view of the prevailing uncertainty, it was found to be extremely important to delegate the control responsibilities as much as possible to the shop floor level. If a separate production phase (i.e., a separate Process Planning PU) were to be created, then this would mean an extra control burden at the GFC level. By delegating the responsibilities for all of the production planning activities (including the ordering of materials and process planning) to the manufacturing and assembly of components, the control at the GFC level needs only be concerned with aggregated information. The concept of a sub order is of primary importance in connection with the delegation of these control responsibilities to the PUs.

A manufacturing sub order, for example, consists of a set of components to be produced. Sub orders are assigned to PUs at the GFC level and arrangements are made with the PUC concerning the internal throughput time for these sub orders. The delivery lead time for ordering materials and the throughput time for carrying out the process planning are taken into account. The sub order represents the lowest level of aggregation for the GFC in connection with controlling a customer order. Sub orders are scheduled based upon the aggregate information. For this reason, the GFC is not interested in detailed information such as process planning data and material requisitions concerning the components within the sub orders. This responsibility belongs to the PUs. In this way the GFC only needs to be concerned about the situation at the sub-order level (see for more details Chapter 21).

22.2.2 Supplementary decision structure for customer driven engineering

Customer driven engineering is performed at the beginning of the total transformation process. Contact is maintained with the market during this phase in the form of receiving requests for quotation, sending quotations to customers and receiving customer orders. More quotations are prepared within the customer driven engineering than the number of customer orders which are received since the success rate of quotations is relatively low. This means that an important objective for the control of the customer driven engineering at the GFC level is to channel the flows of requests and quotations and to make selections in such a way that:

- the resulting customer orders provide a maximum contribution to the realization of the business objectives;
- the company receives a balanced flow of customer orders to be processed, matched to the processing capacity of the production organization and satisfying the conditions which have been established for being able to meet the agreed due dates.

The channelling and selection of potential orders is carried out based upon a large degree of uncertainty, however (see also Chapter 9). The designed decision structure should take this into account by monitoring and repeatedly reviewing the requests currently being processed in the light of the basic assumptions which have been established and the current circumstances. If we examine the decision structure developed for the total transformation process as presented in Figure 21.1 with this in mind, then we arrive at the conclusion that using only the decisions regarding "Customer order acceptance, pricing and due date assignment" and "Sub-order assignment and outsourcing" at the GFC level is insufficient for realizing this objective. In order to be able to achieve the aforementioned objectives at the GFC level, the decision structure needs to be supplemented with three additional control decisions to be added to the two decisions already mentioned. This creates a decision structure consisting of a total of five decisions at the GFC level for controlling the customer driven engineering process:

a. Evaluating and selecting requests for quotation: which of the received requests for quotation will be selected as the basis for preparing quotations?
b. Issuing quotations: which quotations will be sent out and which conditions will be stated regarding the price and delivery time?
c. Reserving resource capacity within the Detail Design PU: will (scarce) resource capacity be reserved ahead of time during the quotation phase for a potential customer order?
d. Internal order acceptance: what price and due date will be quoted for the customer order and what will the consequences be for the production organization?
e. Sub-order assignment and outsourcing: which production unit will be producing which components or assemblies and what part of the work will need to be outsourced?

Figure 22.1 illustrates the positioning of the various decisions within the decision structure for customer driven engineering in relation to the identified control levels which were presented in Figure 21.1.

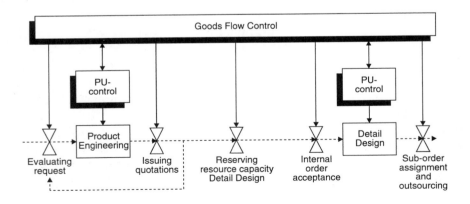

Fig. 22.1 The control diagram for customer driven engineering

Two levels can be defined for these five GFC decisions in view of the different degrees of importance. Decisions regarding "Evaluating and selecting requests for quotation", "Issuing quotations" and "Internal order acceptance" are seen as a shared responsibility of all of the members of the management team due to the possible (integral) impact on the organization. The decisions regarding "Reserving resource capacity within the Detail Design PU" and "Sub-order assignment and outsourcing" are seen as the responsibility of the respective line managers due to the local impact of these decisions. The three supplementary control decisions at the GFC level are described in further detail below, including the reasons why each of these additional decisions is required.

Evaluating and selecting requests for quotation
The purpose of the decision regarding "Evaluating and selecting requests for quotation" is to chose a subset of the requests which are received by Sales in such a way that the quotation preparation capacity which is available within the Product Engineering PU can be optimally utilized. The questions here are which requests should be honored in the sense that a quotation will be prepared, and how much resource capacity should be made available to prepare these quotations. The received requests for quotation are reviewed to determine their desirability in terms of several criteria. These criteria fall into a number of general categories:

- product/market policies; how the request fits in with the product/market policies;
- financial attractiveness;
- the probability that a firm order will result from the quotation;
- strategic potential;
- technical risk; the risk of not being able to produce the product or not being able to produce the product at an acceptable cost level;
- requirement for product engineering capacity to prepare the quotation (related to the technical risk);
- current utilization level of the product engineering capacity;
- the current market situation (positive or negative economic climate).

There are two reasons at this stage for taking little notice of the expected future utilization of the physical production capacity. In the first place, there is a large amount of uncertainty regarding the equipment specifications and regarding the chance of actually winning the order and the actual production start date. This means that any assumptions which could be used to determine the availability of future resource capacity will not be reliable. In the second place, the production organization can always make use of outsourcing to increase its flexibility, if necessary.

As an example, we will describe in further detail how this decision was implemented in the case situation described in Chapter 9.

In the first place, the PMC strategy was defined and translated into operational terms in collaboration with the Management Team. This was essentially a statement about the degree of attractiveness to be assigned to each combination of product group and geographical marketing area (the PMC) and the sales and profit margin targets for each PMC. The received requests could be sorted into the important categories for further consideration based upon the clarification of the PMC strategy and communicating this to the organization. Each PMC was also divided into market sectors (in addition to geographical areas and product groups) primarily for the purpose of setting targets for the Sales organization. Examples of market sectors for bottling machines include the distilled beverage industry (e.g. whisky), the dairy industry and the soft-drink industry. Each combination of PMC and market sector was assigned to one of the following categories:

- an important PMC for which new orders should be actively solicited (Category A);
- a lesser important PMC for which no special marketing efforts are warranted (Category B);
- an unimportant PMC for which requests for quotation generally will not be accepted (Category C).

In the second place, a procedure was developed for a *business evaluation* and a *technical evaluation* for the requests for quotation and subsequently implemented in the selected case situation. The business evaluation is carried out within the Sales organization. The respective salesperson evaluates a request based upon four main criteria, namely:

- the product/market strategy which dictates whether this request falls into a PMC which is important to the business;

- the chance of winning the order, based upon information about which competitors are also preparing quotations and whether this is an existing customer;
- an estimate of the potential revenue from this sale (the price level);
- the follow-on potential of this request for quotation, based upon information about whether this is an important opportunity in an important market segment, whether the customer is the market leader and whether there are possibilities for additional orders.

The business evaluation is carried out by completing a standard business evaluation questionnaire. The requests are discussed once each week within the Sales organization and a decision is made regarding which requests are sufficiently attractive from a business point-of-view. In addition, a normative internal throughput time is assigned to each request to indicate the deadline for completing the quotation. The assignment of a due date in this way is useful for planning the work to be released to the Product Engineering PU. The technical evaluation is carried out within the Product Engineering PU by filling out a standard technical questionnaire. The purpose of this questionnaire is to provide support in estimating the technical risk associated with the request for quotation based upon the functional requirements specification. As the number of customized requirements included in the request increases, there is a greater likelihood that more than the originally estimated product engineering and detailed design hours will be needed, thus increasing the technical risk.

"Evaluating and selecting requests for quotation" takes place weekly at a meeting of the Quotation Team in the case situation. This Quotation Team is comprised of the members of the Management Team, the heads of the Packaging Technology and Bottling Technology sections of Product Engineering and the Materials Manager.

Issuing quotations

After a request has made it through the evaluation and selection stage, it is ready for release to the Product Engineering PU. A preliminary draft of the product is prepared at this point and used as the basis for a quotation. The product description is developed to the extent necessary for establishing an acceptable price and delivery due date with a limited risk. Nevertheless, it is important that a periodic review is carried out at the aggregate control level to evaluate the criteria for the quotations to be issued. The reasons for this review are as follows.

In the first place, the preparation of a quotation may take a significant amount of time. It is possible that, in the meantime, the basic assumptions that were used for selecting and accepting the request have changed. In the second place, the basic assumptions underlying the request could change during the negotiation phase to the extent that the reasons for accepting the request are, similarly, no longer valid. For example, it is possible that the customer has changed the original specifications with the result that the technical risk has become too great in relation to the financial attractiveness. The decision concerning "Issuing quotations" is also made at the meeting of the Quotation Team.

Reserving resource capacity within the Detail Design PU

Reserving resource capacity for potential orders within the Detail Design PU is an important interim control decision in which timing considerations play a significant role. In most cases, the current utilization levels of the resource capacities are not considered in connection with issuing quotations. The Detail Design PU may be an exception, however. In view of the fact that this resource capacity is the critical, limiting capacity during the order phase, a majority of the problems are likely to occur in this area. Under certain circumstances it may be decided to include the influence of certain quotations in the evaluation of the capacity loading within the Detail Design PU. In this way it is possible to determine whether a specified delivery due date is feasible if the customer decides to place the order at a certain point in time. In addition, it is also possible to actually reserve future resource capacity within the Detail Design PU for a potential future order. Since the probability of winning most orders is relatively low, reserving capacity in advance in this way will certainly not be done for every quotation. Any reservations of capacity for potential future orders should be evaluated carefully. By not reserving capacity in advance, delivery problems may occur. On the other hand, reserving capacity which may never be used can lead to an under-utilization of future capacity or a loss of future customer orders. The decision to reserve capacity in advance is made based upon the consensus of Sales, Engineering and Materials Management.

23

Production Activity Control

23.1 INTRODUCTION

We have argued in previous chapters that a distinction should be drawn between Goods Flow Control (GFC) and Production Unit Control (PUC). Goods Flow Control can be broken down into Aggregate Production Planning and Factory Coordination. Production Unit Control is also called Production Activity Control, see Figure 23.1. While GFC has been discussed in previous chapters, this chapter will discuss PUC in more detail. The role of PUC is in principle to execute the plans of Factory Coordination. These plans are materialised in the form of concrete work orders. However, the role of PUC is not merely the execution of these orders, it requires an explicit release of these work orders to the shop floor. In addition, it requires a local planning function and a dispatching function. The goal of this chapter is to provide the control architecture at this level in the factory.

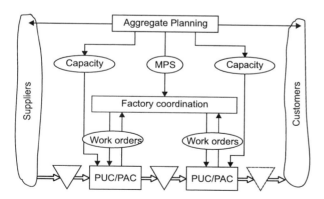

Fig. 23.1 Production Unit Control/Production Activity Control in a wider context

The control architecture at the Production Units level consists essentially of three functions:

1. work load control
2. scheduling
3. dispatching.

Each of these functions can be divided into two major decisions. Work load control can be divided into work order release and work load acceptance. Scheduling can be divided into the allocation of capacity to specific work orders and the detailed planning of work orders. Dispatching can be divided into the dispatching of work orders and material release.

In order to be able to make the work order release decision, the decision maker has to consider both the material aspects and the capacity aspects of the work to be released. Moreover, one should not just consider the orders that can be released immediately. This would eliminate the need for an explicit decision. Work order release will for this reason usually occur in two stages:

1. the Production Unit will have to determine how much capacity is available for new work orders in the following period, using for example techniques for detailed capacity planning. Backlog and finished work will play an important role in this decision.
2. a selection has to be made of orders to be released. This decision will typically be made in cooperation with Factory Coordination.

The work order release decision implies that the Production Unit accepts a specific delivery date for a work order. Often, a detailed time-phased schedule of operations will be established upon release of a work order in which it is indicated how the delivery date will be met. The objective of this is to be able to check whether the Production Unit is on schedule. This schedule will be used when dispatching work and allocating capacity.

Material needed for the execution of a work order is in many cases available upon the start of that work order. In some situations, however, it may be desirable to have the material only physically available upon the dispatch of a particular operation of the work order. There will be a combined dispatching of operations and material in these situations.

This discussion results in the decision structure presented in Figure 23.2. Each of the functions will be discussed in more detail in the following sections.

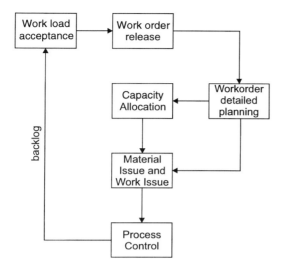

Fig. 23.2 Decision structure for Production Unit Control

23.2 WORK LOAD CONTROL

From the viewpoint of the Production Unit, it only makes sense to release a work order with a specific delivery date to the Production Unit if the Production Unit can indeed guarantee that the order will be delivered at the specified delivery date. It would be better not to release this work order if delivery were to be impossible for capacity reasons. Why would this be better?

Firstly, the refusal to release work orders because of capacity restrictions makes it clear that there is a problem. There is only one solution for the Production Unit to solve that problem and that is to increase its capacity. Only then will it be possible to release the particular orders without endangering the earlier agreements made on delivery dates.

Secondly, in most cases there are alternatives available at the Factory Coordination level. For example, the safety margins (such as safety stocks and slack time) further downstream can be used. Negotiations can be started with the customer. One can try to modify batch sizes or realise throughput time reduction further downstream in the production line.

Thirdly, priorities concerning critical capacities can be re-specified at the Factory Coordination level in cooperation with Marketing and Sales.

As such, one should consider any order releases made by a Production Control system indeed as a suggestion. There are various reasons why the actual release of work orders from Factory Coordination to the Production Units may be different from the suggestion, the most important reason usually being the availability of capacity at the Production Unit level. It is for this reason that we suggest including an explicit Work Load Acceptance function in the diagram in Figure 23.2. This function determines how much new work in various categories can be accepted, based on the actual shop floor situation. Any disturbances on the shop floor will immediately lead to fewer orders being accepted. Any windfalls in production will on the other hand be used to increase order release.

The way to implement work load acceptance in practice varies. The three best-known techniques are:

- Gantt charts;
- finite capacity planning
- work load control.

We will discuss each of these techniques below. In passing, we will discuss work order release procedures.

Gantt charts
Gantt charts can be used both for work order release to dispatching (which will be discussed in more detail in section 23.4). This means that a very detailed overview is kept of all capacities, their planned activities and their sequences. Upon release of a new work order, this plan is reconsidered in such a way that the order can be realised with available capacity. Attempts will thereby often be made not to shift existing, planned orders. The requirement for this technique is that there should be few disturbances over the planning horizon, that feedback from actual progress of work orders on the shop floor should be processed immediately in the Gantt chart, and that the capacity should be flexible enough to eliminate the remaining disturbances.

Finite capacity planning

In finite capacity planning, all capacity types are considered in terms of fixed periods of, for example, one week. More precisely, sequence relations between the various activities within a period cannot be maintained. Initially, all capacities in a period are filled up with the orders already released, based on planned delivery dates. In the process, capacity restrictions or past-due dates will be ignored. This will result in a diagram like the upper diagram of Figure 23.3. After that, the past-due orders are rescheduled to the first available location in the future. Thereupon, the overload is eliminated, preferably by moving a 'top' to an earlier time (Figure 23.3, lower part). Notice, however, that shifting capacity can be quite questionable, especially when one would think how the result would reflect reality. After all, interactions between different diagrams are disregarded. These interactions do exist since the same orders will be processed on various capacities. The counter-argument would be that this would not make a difference for the acceptance of future work. Bertrand and Wortmann (1981) consider these arguments in more detail and criticise this method of capacity planning. Their analysis leads to Work Load Control.

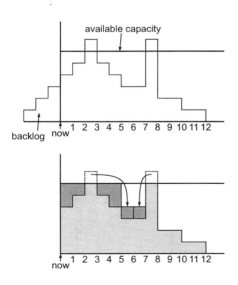

Fig. 23.3 Finite capacity planning

Work load control

In a number of situations with a steady pattern of work order release, it will be sufficient to keep the shaded area in Figure 23.3 (lower part) constant. This is the principle of work load control (see Bertrand et al. (1990)), which is closely related to the idea of Input/Output control from MRP-II (see Figure 23.4).

The advantage of work load control is that it is necessary to consider only one figure per capacity type in order to realise order acceptance successfully. Work load control does, however, presuppose control of a few preconditions. It also presupposes a certain level of training and knowledge in management and planning. The mix of released routings especially should not divert too much from the mix agreed upon in aggregate planning.

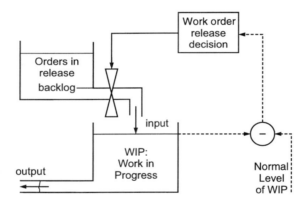

Fig. 23.4 Work load control

23.3 SCHEDULING: CAPACITY ALLOCATION AND DETAILED WORK ORDER PLANNING

23.3.1 Capacity allocation

Capacity allocation is concerned with specifying available capacity. Essentially, there are two forms of capacity allocation: (1) based on multi-task capability of machines and/or operators, or (2) based on increasing or decreasing capacity in the short term.

Multi-task capabilities
The first form of capacity allocation concerns the allocation of capacity to various 'types of work'. Particularly when considering human capacity, it will often be the case that people are multi skilled and can do various types of work. Multi-skilled workers are an important source of flexibility. It enables management to respond to disturbances such as illness, change in personnel, quality problems, etc. Moreover, in most factories, job-shop-style and other, there will be more machines available than operators. This means that the bottleneck in these factories will seldom be created by machines, but rather by humans. Obviously, one of the most important daily decisions of management will then be to decide who is going to work on which machine. Multi-skilled labour does not, however, emerge automatically, but has to be managed explicitly. It requires a dedicated training programme. Moreover, the workers should be rotated to preserve the multi-skilled attributes of the workforce obtained. It will also be beneficial in that this will often raise the quality of work at the same time.

Increase/decrease of capacity in the short term
In many factories, it is possible to increase or decrease capacity in the short term for a period of a couple of days. It is important that the term for increase/decrease should be shorter than the throughput time of the orders since the decision is meant to increase the product flow of the released work orders. Capacity increase is accomplished by means of:

- overtime;
- contracting out work;
- moving maintenance work or training;
- splitting up production batches to increase the product flow towards following machining centres;
- combining set-up times to increase capacity at a bottleneck
- etc.

Each of these decisions puts certain requirements on the information system, whether manual or computerised, which should indicate both problems and possible solutions. Gantt charts can be used here to calculate the consequences of certain decisions.

23.3.2 Detailed work order planning

A method often used to support capacity allocation is to make a detailed work order schedule for all work orders. This involves the assignment of a due date/time to each operation in the routing of a work order. This due date/time is called the planned date of the operation. Normally, the planned date of the last operation should equal the delivery due date of the work order (disregarding safety margins and transport time).

The difficulty in assigning planned dates is the estimation of the expected waiting times of the order. These waiting times are dependent on the utilisation level of the work centre, which in turn is dependent on future allocation decisions. These mutual dependencies of waiting times and utilisation levels are the key control problem in Production Unit Control. To avoid this circular relation between waiting times and utilisation levels, it is usually assumed possible to guarantee that the delivery times of the released orders can be met. This will be accomplished using the techniques described in the previous section of this chapter, which imply that the utilisation level for each work centre will not exceed a certain maximum. The average planned waiting time will then correspond to this maximal utilisation level.

23.4 DISPATCHING

Dispatching of work order operations and material are the final decisions that have to be taken before operations can start. To begin with, we shall provide a few definitions:

- *Dispatching of work order operations*. This is the decision on the next job that will be carried out by a specific capacity. A 'job' usually consists of an operation to a work order. Often, the work orders will be waiting in a queue before the specific capacity. Naturally, this queue does not need to exist in a physical sense, but can exist in the computerised or other information system, which will be discussed in more detail in chapter 30. The capacity in question can

either be a machine or a human capacity or a combination of man and machine.
- *Dispatching of material*. This is the decision to make the material, information and tools necessary to perform an operation available. In practice, it involves all the requirements not performed by the capacity allocation.

Neither function will always be made explicit: dispatching of work orders will often be lacking in a company with a product-oriented layout since capacities are set up in such a way that there will be little choice as to what to dispatch. In a resource-oriented job shop, making parts, on the other hand, there may be no explicit material dispatching since the material will be made available at the time the work order is released to the Production Unit. In general, however, it is wise to consider both functions, even if one of the two may be absent.

Dispatching of material
In the remainder of this section we will focus mainly on the dispatching of work orders. Different methods for dispatching material are mainly of importance to the design of the information system. We will limit our discussion here to a short overview of the dispatching of material.

A distinction can be drawn between the dispatching of material per order or per operation. In the case of dispatching per order, all material requirements will be dispatched upon the work order release. The advantage of this approach is that one can be sure of the availability of material. The disadvantage is that material and documents will be made available too early, while tools may be available long before the operations will be performed. These pros and cons are reversed in the case of dispatching material per work order operation.

This discussion will make clear why the dispatching of work orders and the dispatching of material are so closely related. When dispatching material per operation, the dispatching of work order operations should not only consider the availability of the capacity, but also the availability of other requirements (in particular materials). This will result in a complicated work order operation dispatch procedure. In practice, preference will often be given to combining the dispatching of material with the dispatching of work orders.

Material dispatching can be achieved by the 'push' system or the 'pull' system. According to the push system, the material required is calculated per work order or operation and the material issued accordingly. With the pull system, the exact material required is used, but a stock of available material is also provided. It is often argued that the push system is mandatory in traditional production planning and control (PPC) systems, while the pull system can be identified with JIT production. This argument is not necessarily true. Nowadays, there are PPC systems available that support a pull dispatching of material.

In the following we will limit ourselves to the dispatching of work order operations and we will disregard the dispatching of material.

Dispatch of work order operations through planning: Gantt charts

The advantage of planning to support work order dispatching is that one can foresee problems with regard to capacity utilisation. It does, of course, require a sufficiently detailed plan to indicate each future dispatch in full detail. This is often realised by visualising all operations per capacity. A plan of this kind is called a Gantt chart (see Figure 23.5). The bars in a Gantt chart represent the capacities, while the horizontal axis represents the time horizon. Different operations on one work order are often indicated by a colour. Traditionally, Gantt charts were large boards covering the walls of an office. There are now computer applications available to support this. These applications are often based on optimisation algorithms or simulation (Adelsberger and Kanet, 1991).

In algorithms for optimisation, one should distinguish static, deterministic algorithms and dynamic, non-deterministic algorithms. The former has the broadest background in literature, and is based on the problem of work orders with capacity requirements and delivery dates being available. Given the available capacity, the problem is to create a Gantt chart that optimises the use of capacities. The problem with these algorithms is that reality often has a number of elements that will not be of a static or deterministic character. Thus, the value of an optimising algorithm is greatly reduced. In reality, the capacity requirements are often based on rough calculations, and exact sequencing does not make much sense. In other words, the detailed level of a Gantt chart only makes sense if the available capacity and required work are known very precisely. The use of a Gantt chart is therefore not recommended for decisions on a long term horizon, but only for use as a supporting technique for the dispatching of work orders. The validity of a Gantt chart is often limited to a short horizon because of the non-deterministic character of the reality. Disturbances concerning the availability of material, availability of capacity, quality, production time, etc., will affect the planning on the Gantt chart. An interactive

Gantt chart used for dispatching on a continuous basis will provide a solution to many of these problems, provided that it is maintained accurately (Jackson and Browne, 1989).

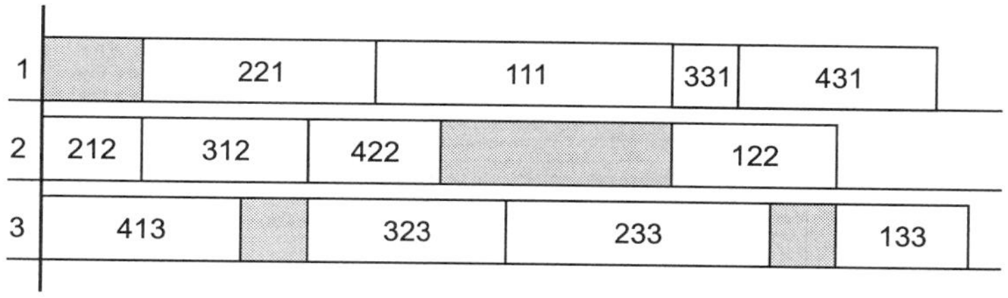

```
1 |    | 221      | 111      | 331 | 431 |
2 | 212| 312 | 422 |         | 122 |     |
3 | 413    |   | 323 | 233       |   | 133|
```

Fig. 23.5 Example of a Gantt chart

Dispatch of work orders through priority rules

The second way to dispatch work orders is based on priority rules. A priority rule provides a guideline as to what job to take from a queue of waiting jobs at the moment the capacity becomes available. Such a rule will thus not determine the sequence for the jobs in the queue, but will only select the one to be taken into operation.

There is extensive literature available on priority rules, especially in the area of job shops. The four basic rules are:

- First In First Out (FIFO);
- Shortest Processing Time (SPT);
- Work-In-Next-Queue (WINQ);
- Earliest due date (EDD).

When comparing these rules, the following criteria will be used:

- the average throughput time of the orders;
- the average past-due time of the orders;
- the variance in late and early orders.

The general idea found in literature on priority rules is the fact that they will not provide a solution to all problems. In other words, priority rules are very suitable for solving relatively easy, repetitive problems of a more or less identical nature. Usually, EDD will be used in combination with one of the other rules.

Simulation

A combination of both methods (planning and priority rules) will often be used in practice by making a Gantt chart in which the future in the short term is simulated by means of priority rules. The computer will thus simulate the near future on the basis of priority rules. It is, of course, impossible to predict any future disturbances. However, future problems that would arise without disturbances can be identified. These simulations will typically provide a view for the next two or three days. The result of a simulation may therefore constitute a reason for replanning, or for making other arrangements, such as overtime, rescheduling maintenance, etc. (which are essentially decisions related to the scheduling functions discussed in section 23.3).

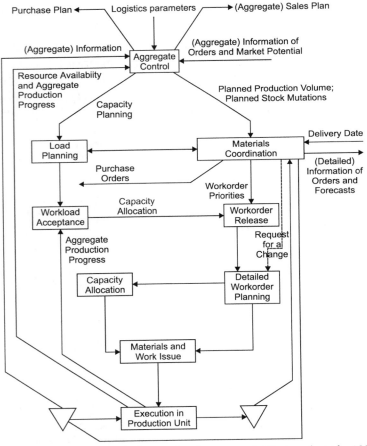

Fig. 23.6 Framework for production control (Bertrand et al., 1990)

23.5 CONCLUSIONS

It should be noted that the diagram in Figure 23.2 presents a considerable extension to the diagram traditionally used in MRP-II. This is not surprising. The role of capacities in shop floor control, especially in Customer Driven Manufacturing, is considerable. It is to a large extent disregarded in MRP-II, however. We therefore ended up more or less as a matter of course with a broader framework than MRP-II. This framework is presented in Figure 23.6.

One issue in Figure 23.6 has not yet been discussed. The figure indicates that capacity utilisation planning in the figure is considered as a more important function than the corresponding detailed capacity planning in MRP-II. The reason is that detailed capacity planning in MRP-II is based on MRP-I, without however creating a relationship to aggregate planning or production unit control. Bertrand et al. (1990), however, argue for a strong relationship between utilisation planning and aggregate production planning on the one hand, and utilisation planning and work load control on the other.

REFERENCES

1. Adelsberger, H.H., and Kanet, J.J., "The Leitstand - a new tool for Computer Integrated Manufacturing", *Production & Inventory Management*, vol. 32, no. 1, 43-48, 1991
2. Bertrand, J.W.M. and J.C. Wortmann, *Production Control and Information Systems for Component Manufacturing Shops*, Elsevier, 1981.
3. Bertrand, J.W.M. et al., *Production Control, A Structural and Design Oriented Approach*, Elsevier, 1990.
4. Jackson, S., Browne, J., "An Interactive Scheduler for Production Activity Control", *Int. J. Computer Integrated Manufacturing*, vol. 2, no. 1, 2-15, 1989.

24

Customer and supplier relations

24.1 INTRODUCTION

With the worldwide increase in production capacities and the emerging global availability of production capabilities, the success factors for all types of industrial production are changing due to intensive competition. Besides improved and effective internal order processing, intensive cooperation with external partners (both customers and potential competitors, suppliers, subcontractors, etc.) is counted among the basic requirements of modern production systems. Although the economic potential contained in such cooperative approaches has been sufficiently known for years, these possibilities have often been used in a non-systematic and/or inadequate way, even in customer order driven manufacturing.

24.2 MANUFACTURER-CUSTOMER RELATIONS

Today, medium- and long-term strategic manufacturer-customer relations in the capital goods industry should span the whole product life cycle, from first customer contact through recycling of the product right up to its substitution by a new product. The principles of manufacturer-customer relations throughout the product life cycle are illustrated in figure 24.1.

Maintaining communication not only internally (within the company itself at all levels and throughout all areas) but also externally is consequently one of the core demands of total customer orientation. Since product-related advantages are being copied by competitors more and more quickly, the customer services accompanying an offered product, from the financial concept to turn-key installation and maintenance of the product, are increasingly regarded by the customer as the differentiation criteria. Consequently, the more the product is

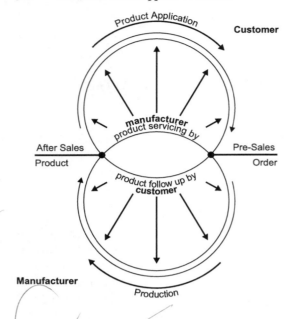

Fig. 24.1 Classification of manufacturer-customer relations in the product life cycle.

tailored to a customer, the more intensive and continuous the contact with the customer should be. Major principles of the most relevant product life-cycle phases are discussed below.

24.2.1 Pre-sales: strategic customer acquisition

A general innovation thrust, microelectronics-triggered for example, is not on the horizon. Therefore, products in the future will distinguish themselves less and less by different technology. Consequently, many enterprises need to pursue new, better-directed strategies of customer service and support during pre-sales. A great number of actions which take into account the above-mentioned requirements in quite different ways are well-known, from practical experience as well as from the literature. The common goal of all these measures is, ultimately, an increase in the productivity of pre-sales activities.

There are, of course, indirect rationalising approaches, i.e. time and cost optimisation during customer cultivation and bid preparation, which are invisible to the customer. But there are also direct customer-oriented actions which - with the active inclusion of the customer - are directed at increasing the probability of getting the contract, at minimising risks during product specification and at expanding the company's business activities.

If the Simultaneous Engineering (SE) philosophy and the active customer inclusion associated with SE are already used during the pre-sales phase, it is not only possible to prevent over-engineering based on misunderstandings in the bid specification but it could also result in increased odds of a new contact. Moreover, a customer who has already invested a great deal of time in the product-specific implementation of his requirements during bid preparation will identify stronger with the product. Such a customer will not be so inclined to withdraw or cancel the contract.

Furthermore, in the capital goods industry the customer's purchase decisions are accompanied, directly and indirectly, by a strong feeling of risk. Questions such as:

- how these capital goods can be integrated into the customer's current application, and
- how quickly the processing capability can be guaranteed

are factors which have to be recognised in advance and for which solutions have to be found. Skilful support given to the customer, thus minimising his feeling of risk, contributes both to the success of order acquisition and to the customer relationship.

As there are no dedicated methods available for customer driven manufacturing, known methods and concepts from the product planning phase of mass production should be adapted to meet the requirements of customer driven manufacturing. Such methods comprise, for example, computer-supported product simulation and animation, QFD (Quality Function Deployment), conjoint analysis, etc.

Consequently, in some cases in the capital goods industry, computer-based simulation models, enabling a most extensive simulation of product behaviour, have been a part of the bid (e.g. complex assembly lines). In principle, it may be assumed that through the increasing globalisation and opening up of markets the number of bids to be prepared by enterprises will increase, although with an increase in bidders a decrease in the realisation ratio can be assumed.

24.2.2 Customer relations during order processing.

This book emphasises that manufacturing is turning away from the anonymous market and towards the individual customer order as the driving force. If so, the product-related knowledge normally obtained during the product planning phase of a prepared mass or series production will be handicapped more and more by heavily reduced possibilities of foresighted product planning. In customer driven production this knowledge has to be gained by early customer orientation during order processing. One of the biggest challenges in customer driven manufacturing is correctly transforming the customer's wishes into product features. In the course of the first steps of the bid preparation phase, throughout which only rough product specifications still prevail, many details have to be decided. The process proceeds, even after the contract has been awarded.

According to the philosophy of Concurrent Engineering the customer should be linked to the supplier by consequent, systematic communication in such a way that late and massive changes can no longer occur. Small changes can and should be discussed between the customer and the personnel responsible for order processing. Immediate response should be possible. Therefore, endeavours should be made to concentrate on critical product functions during the early stages of the product life cycle. In this case, based on experience gained from previous orders, product characteristics which are difficult to change must be differentiated from those which are easily adaptable in the short-term.

At the same time, this requirement assumes the willingness and determination of the manufacturer to give the customer access to project progress during order processing as well. Often, the existing process organisation has to be reviewed for customer suitability. Accordingly, new customer communication channels have to be created. The goal has to be the pursuit of preventive quality assurance through a maximum of lucidity on the part of the customer to bring about a maximum of customer satisfaction through the product to be manufactured.

To enable all participants needed for product specifications to work as a team and to bring together all information required, systems are needed which not only allow the transmission of data (audio and video), but also integrate it into one common user interface. Such a system was developed within the project DIMUN (Distributed International Manufacturing Using Existing and Developing Public Networks),

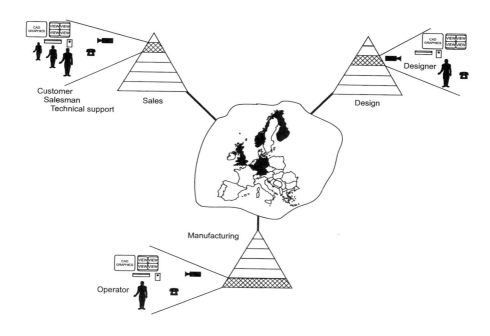

Fig. 24.2 Customer-integrated, globally distributed order processing

whose aim was to identify the potential uses of broadband communications for customer integration in the globally distributed order processing of capital goods [Leev-91].

The DIMUN multimedia manufacturing demonstrator (figure 24.2) shows how customers might interact during order processing with a future transnational manufacturing enterprise. The system developed offers distributed multimedia control and communications combined with simple Computer Supported Cooperative Work (CSCW) functions to support interacting partners working together in remote locations. Each partner works within a "room" (user-interface) with software which allows the partners to exchange documents and to work together on a shared "whiteboard" offering sketching facilities. Meeting members can operate on different parts of the whiteboard at the same time. The changes are broadcast to all meeting participants.

Each partner's pointer is also shown on the other screens. In addition, the partners can see each other on a "tiled" video display in another window on the screen. A Uniplex multiview unit is used in the demonstrator to simulate future, more flexible video windowing. The Uniplex presents up to 16 views under computer control in a wide variety of tiled window layouts.

24.2.3 After-sales customer service

After the production phase has finished, the product application phase begins with delivery and putting the product into operation at the customer's facilities (figure 24.1). Nowadays, the additional business possibilities resulting for the manufacturer from "life-long" customer service should be emphasised in the interests of total customer service. With the trend towards shifting added value from the production phase to the aftersales phase, the sales obtainable in the after-sales area often lie significantly above the original amount of the contract, depending on the product. This fact is a discovery which is only gradually being realised by many manufacturers (for example, extensions, maintenance and repair work for complex products like planes, ships and plants). In times of increasingly intensive competition, a business sector can thus be developed for a manufacturer in which he alone is regarded as the contractor.

Besides other approaches, concepts allowing the global maintenance of product service and support are necessary. Consider, for example, the MARIN-ABC project (Marine Industry Applications for Broadband Communication) which focused on increasing the safety and efficiency of global shipping operations [Schw-92]. Specifically, it demonstrated how a non-routine repair problem can be solved in accordance with customer wishes more efficiently and economically through the "via satellite" assistance of shore-based experts using the advantages of multimedia communications. The test platform, which was similar to the platform of the DIMUN project, consisted of applications for high quality voice, live video, high definition still photography, fast CAD-data file transfer, interactive data transmission and real-time sharing of computer applications between a railway ferry and several shore-based experts. The communications infrastructure developed can be applied to other uses, such as the support of remote repair, installation or maintenance processes.

24.3 MANUFACTURER, SUPPLIER AND SUBCONTRACTOR RELATIONS

Due to the recent developments in the automotive industry, interorganisational cooperation is at present mainly discussed in the context of mass and serial production. However, this cooperation also offers a good basis for future developments in all types of customer driven manufacturing. A company that develops more quickly than its competitors not only has the possibility of starting later, but it can also respond with greater flexibility to changing customer requirements. If internal and external capabilities for development and manufacturing can be brought together, cost benefits will be achieved, because specialist partners may offer different and beneficial cost structures.

Corresponding to the large variety of customer-specific products, various kinds of cooperation exist between enterprises today. Modifications or new developments require a high proportion of previous solutions to be integrated, whereby a significant percentage of parts have to be purchased. For example, in packaging machine design 70% of material costs are based on parts purchased, and for assembly lines the rate is 90% or more even [BIRK-92]. Accordingly, the majority of customer order driven manufacturers have long, but extremely different, experience in cooperating with suppliers and subcontractors. Nevertheless, a systematic strategy for "lean product development", "lean design" or "lean manufacturing" for customer-specific products is lacking. Although a systematic reappraisal to produce an overall strategy for integrating external expertise into order processing in customer order driven manufacturing is beyond the scope of this contribution, some approaches can be discussed.

Depending on the level of integration of a particular type of cooperation and the phase of the product life cycle, various cooperation strategies can be established by concentrating on the manufacturer's core activities. These core activities will differ from company to company (figure 24.3). Some of these approaches are described below.

24.3.1 Cooperation in product development: reduction in development scope

Cooperation between enterprises aiming at the strategic procurement of additional competences is unusual in customer driven manufacturing (apart from specific cooperation in large project-related consortia).

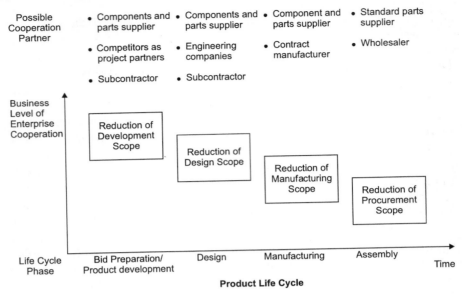

Fig. 24.3 Possible approaches to producer-supplier cooperation in customer order driven manufacturing.

However, a focused configuration of enterprises (suppliers, subcontractors, etc.) temporarily cooperating in a harmonised way could present a substantial solution for future customer driven manufacturing. Close cooperation between independent enterprises offers the possibility of coping with increasing customer and market requirements in regard to product functionality and quality. The targeted combination of internal and external cost and time advantages offers an exceptional economic potential which cannot be achieved by a single enterprise with a "do-it-yourself" strategy. Cooperation in product development is a chance for the manufacturer to concentrate his own development activities on achieving various complementary product designs. At the same time the capability of integrating available components and subsystems into an overall system should be seen as an essential core competence of the producer [Bout-93].

Whereas the implementation of "total customer proximity" requires new organisational solutions, the successful integration of external partners places additional demands on the application of information and communication technologies. Furthermore, the cost, time and quality advantages mentioned above can only be achieved if the coordination and transaction costs are minimised. The distribution of development activities to achieve parallel processing of subprocesses leads to increased risks which must be met by optimised and harmonised information management. The definition of clear release procedures to support the pre-release of subresults will become an essential element in accelerating the processing of customer orders.

A concept has been developed for the system-based coordination of intra- and interorganisational product development activities within the ESPRIT project CMSO (Cim for Multi-Supplier Operations) which allows information exchange in manufacturer-supplier cooperation independent of hard and software applications [Lisc-92]. Using the most modern communication systems and technology the "Technical Information Management System (TIMS)" was developed which, in particular, supports data exchange during product development. Within the TIMS prototype system technical data is directly exchanged via neutral interfaces like IGES VDA-FS and VDA-IS. The economic data format Odette is used for order and quality-related data.

24.3.2 Cooperation in design: reduction in design scope ("Design or Buy")

Even if cooperation in product development is not appropriate, cost and time reductions can be achieved by reducing activities in the design area. The importance of purchased components and parts for product development is offset by a number of weak points in a systematic integration of existing (potential purchased parts) or known parts (former designs) and components during product development and design. Up to now the systematic retrieval, identification and integration of these parts and components into a new product has only been very roughly supported by design methodologies, design concepts and computerised systems [BIRK-92]. Increased reuse of known design solutions and the prevention of unnecessary new designs by using purchased parts and components represent - in addition to organisational approaches - the basis for future productivity increases in design departments. Figure 24.4 gives an example of possible benefits when choosing a bearing unit as a purchased component instead of developing a design of one's own.

Convincing the design and development departments that an increase in the use of purchased parts and components would mean substantial economic benefits compared with new and often risky

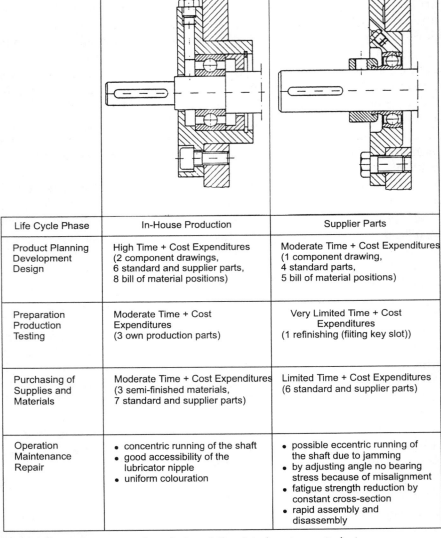

Fig. 24.4 Comparison of "make" and "buy" in bearing unit design

Life Cycle Phase	In-House Production	Supplier Parts
Product Planning Development Design	High Time + Cost Expenditures (2 component drawings, 6 standard and supplier parts, 8 bill of material positions)	Moderate Time + Cost Expenditures (1 component drawing, 4 standard parts, 5 bill of material positions)
Preparation Production Testing	Moderate Time + Cost Expenditures (3 own production parts)	Very Limited Time + Cost Expenditures (1 refinishing (fiiting key slot))
Purchasing of Supplies and Materials	Moderate Time + Cost Expenditures (3 semi-finished materials, 7 standard and supplier parts)	Limited Time + Cost Expenditures (6 standard and supplier parts)
Operation Maintenance Repair	• concentric running of the shaft • good accessibility of the lubricator nipple • uniform colouration	• possible eccentric running of the shaft due to jamming • by adjusting angle no bearing stress because of misalignment • fatigue strength reduction by constant cross-section • rapid assembly and disassembly

design solutions is becoming an essential task for design and development managers. While relevant decisions, global contexts and enterprise strategies are not sufficiently understood, design and development staff cannot be expected to change their way of thinking.

Additional possibilities for reducing the design scope are presented by subcontracting adaptive design and drawing activities. Subcontracting entire area-related design activities (e.g. pneumatic and hydraulic applications in machine tool design) represents a starting point. The prime characteristic of such cooperation is a detailed specification of requirements demanded by the manufacturer as a basic prerequisite for successfully integrating a subcontractor. These specifications often do not emerge if the "peripheral" design activities thus considered are executed as a side-line of overall internal design activities.

24.4 SUMMARY

Future company success will depend to a great extent on how far companies succeed in involving customers, suppliers, subcontractors etc. as active partners in the production process. This involvement includes both organisational and IT issues, in the context of Concurrent Engineering for the entire product life cycle. Manufacturers need to focus all activities, from pre-sales right up to after-sales service, on customer service and support for the entire life cycle. A significant difference from the present understanding of the customer-manufacturer relationship will be the development of a cooperative, product-life-cycle-oriented partnership between customer and manufacturer, based on mutual trust. On the other hand, the capability of offering customer-specific products and services at competitive prices and within a suitable time frame will depend more on the enterprise's potential to integrate internal and external expertise and capacities. In addition to enabling the integration of the customer into the production process, this potential for integrating suppliers is becoming a major prerequisite for all types of customer driven manufacturing.

REFERENCES

1. [Birk-92] Birkhofer, H
   Erfolgreiche Produktenwicklung mit Zulieferkomponenten; in: Praxiserprobte Methoden erfolgreicher Produktenwicklung; VDI-Berichte 953, 1992
2. [Bout-93] Boutellier R, et al.
   Produktenwicklung; in: Wettbewerbsfaktor Produktionstechnik -

Aachener Perspecktiven; Herausgeber AWK, Pfeifer, T. et al.; VDI-Verlag, Düsseldorf 1993
3. [Leev-91] Leevers, D., Condon, C., Lutz-Kunisch, B. and Ahlers, R.
The Dimun Project: Experiences with Multimedia Communications in "One-of-a-Kind" Manufacturing; Proceedings of the IIP TC 5/WG 5.7 International Working Conference on New Approaches Towards "One-of-a-Kind" Production; Editors Hirsch/Thoben; Bremen, Germany, 12-14 November 1991
4. [Lisc-92] Lischke, C.
Abstimmung und Koordinierung von unternehmensubergreifenden Prozessen der Produktentwicklung; Dissertation Universität Bremen, 1992
5. [Schw-92] Schwantke, G.
Multimediale Unterstützung von Wartungs- und Reparaturaufgaben; in: Konferenzbericht TELEMATICS'92, ISL, Bremen, 1992
6. [SEBA-91] Sebastian, K.-H.
Innovationsmanagement - Die Überwindung von Marktwiderständen; in: Der Vertrieb von Investitionsgütern im europäischen Wachstumsmarkt; VDI-Berichte 889; S. 109-125
7. [THO-90] Varnholt, R.
Welchen Nutzen erwartet der Kunde vom Vertrieb?; in: Erfolgreicher Verkaufen durch systematische Marktbearbeitung; VDI-Berichte 954, 1992, S.21-33

Part E
INFORMATION TECHNOLOGY

25

Introduction to Part E: Information Technology

Information technology (IT) plays a key role in the operations and management of customer driven manufacturing. Roughly speaking, IT can be applied to support:

- product documentation
- resources management
- workflow management
- decision making.

Note that the three views on manufacturing organisations introduced in Chapter 5 are reflected in the last three points.

Product documentation comprises both data management and document management. *Data management* refers to the content of documents. It has to make sure that the content of a document satisfies the specifications and the quality standards of an engineering discipline and of the organisation itself. *Document management* refers to procedural aspects like change management, workflow management and reuse management independently of the content of a document.

The next chapter, Chapter 26, is devoted to product data management. More specifically, *generic* product data management is the topic to be discussed, in succession to Chapter 16.

Chapter 28 is devoted to document management in customer driven manufacturing in general.

IT support of manufacturing is first of all discussed in the context of *standard software packages for business support*. Sometimes these packages are called Enterprise Resources Planning (ERP) software. These packages are discussed Chapter 27. We will describe classical packages for manufacturing standard products (MRP-packages) and show the contrast with information systems for customer-driven manufacturing. During this discussion it will be demonstrated that the above-mentioned views are always present. More specifically, product-data, workflow-data, and resources-data will be described by data-structure diagrams. Decision support is represented by functionality of software.

Subsequently, IT-support of engineering management in customer-driven systems is given in Chapter 29. It will turn out, that customer driven engineering is difficult to grasp because of the fact that the results of all activities are documents, and not physical products.

Finally, IT support of shop floor control in customer-driven systems is specified in Chapter 30. The differences between customer-driven shop floor control and shop floor control in anonymous production are mostly related to uncertainty. Chapter 30 does not so much show differences in the software architecture, but it shows differences in the role of scheduling and planning. More specifically, planning systems in a customer-driven environment should provide facilities for:

- integration between rough planning and detailed planning
- planning based on rough calculations for work orders
- replanning based on large deviations between estimated operation times and realized times.

26

Generic Product Modelling & Information Technology

26.1 INTRODUCTION

This chapter can be seen as an extension to Chapter 16, Generic Product Modelling, in which a product model for specifying a range of product variants was presented. The purpose of this chapter is to explain the generic bill-of-material approach in more detail by describing how the data base is structured. More specifically, we will focus on the data model used to structure the information which is captured in the data base. Less attention is paid here to the generic bill-of-material application which uses this data model.

A brief summary of other product modelling approaches was presented in Chapter 16. Although we acknowledge the importance of other types of product models, we will not discuss these models and the associated information technology implications here. Where possible, we will point out some of the apparent similarities with the parametrised CAD approach. This chapter focuses primarily on the use of generic bills of material during the product development phase. Major decisions are taken during this phase which affect the operational use of product family data by sales, manufacturing and logistics personnel and systems. Using a data model, some general design rules are formulated here which could be used to simplify development activities and to incorporate features which address product life cycle issues at an early stage.

26.2 DATA MODEL FOR GENERIC BILLS OF MATERIAL

The structure of data entities and their relationships can be expressed in the form of a data model [Griethuysen, 1982]. Such a data model can be used to describe the generic bill-of-material approach in more detail. In the generic bill-of-material approach, three different points of view can be identified. The first two are defined independently of the customer orders; the third point of view is then required to define customer dependent product variants. These three points of view can be described as follows:

- the customer's point of view (parameters and parameter values);
- the product family point of view (generic products); and
- the product variant point of view (specific product variants).

Several data entities are required with respect to each of these points of view for storing, retrieving and manipulating information about the product family. The data model in Figure 26.1 shows how these entities are interrelated.

The data entities are described in more detail in the remainder of this section. An explanation of the generic bill-of-material approach was presented in Chapter 16.

The customer's point of view

Each parameter related to this point of view has one or more parameter values and influences one or more generic products. Parameter values may be restricted by a configuration constraint to prohibit unacceptable product variants. The reasons for imposing configuration constraints may be due to marketing strategies (sales constraints), engineering capabilities (technical constraints), assembly restrictions, service requirements or even governmental legislation.

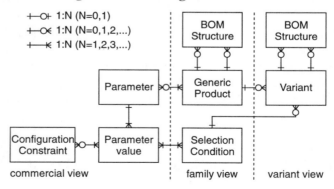

Fig. 26.1 Data model

The product family point of view

Each generic product is associated with zero or more parameters and has zero or more variants. Furthermore, generic products may be comprised of component generic products and be included as components of parent generic products. The structure of a generic product is defined in the bill-of-material (BOM) entity. The selection conditions for determining a specific product variant are related to the customer's point of view as well as the product variant point of view. Nevertheless, this data is normally maintained as customer order independent information which belongs to the product family point of view.

The selection conditions serve to link the customer's point of view with the product variant point of view and are maintained independently from the customer orders.

The product variant point of view

Each product variant is associated with a generic product and is either assembled from other product variants or selected based upon the selection conditions. The selection conditions are only used for primary variants; the secondary variants are always assembled from other primary and secondary variants. Variant BOMs are used for manufacturing (the "as built" situation) and configuration management (the "as maintained" situation).

Variant BOMs are stored in a manner which is similar to that of the product family structure. There is an entity for the variants and an entity for the variant relationships. The set of variants can be divided into two subsets: primary variants (normally customer order independent) and secondary variants (assembled on customer order from primary variants and other secondary variants). Figure 26.2 shows an example of a BOM structure for a product variant.

In this example, the customer order decoupling point [Hoekstra,

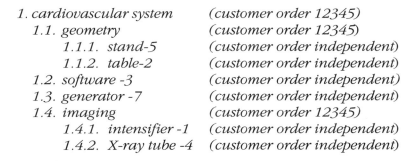

Fig. 26.2 Variant bill-of-material

1992] occurs at the level of the primary variants: all primary variants are manufactured based upon a production plan while all secondary variants are assembled to customer order (12345). From this example we can conclude that the product variant structure is identical to the structure of the product family which consists of GSPs and GPPs [Hegge, 1991]. However, in our data model we will represent the product variant and the product family as separate structures since the structure of a particular product variant may change during the life of the product due to servicing and upgrading. After manufacturing, maintaining the product variant structure becomes the responsibility of Medicom's configuration management department.

In the next section, we will briefly explain how a generic bill-of-material application uses the data model. The possible integration of this with other manufacturing systems such as sales and production control applications is not covered here.

26.3 GENERIC BILL-OF-MATERIAL APPLICATION

A generic bill-of-material application is normally structured based upon the three different points of view. A "configurator" based upon the customer's point of view is generally available to define a valid product variant in terms of parameters and parameter values. A "generator" based upon the product family point of view can then be used to generate a variant bill-of-material from the product family structure. Variant bills-of-material can then be used from a product variant point of view for manufacturing and, in a later stage, for servicing. Figure 26.3 illustrates this.

Although the configurator and generator are targeted for different groups of users, there is a considerable overlap of information. The main entities for the configuration process are the parameters, parameter values and constraints. Since the parameters are linked to generic products, however, the generator application also uses the structure

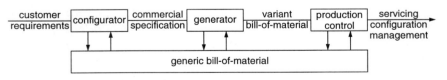

Fig. 26.3 Generic bill-of-material application

Generic bill-of-material application

of the generic products, i.e. the BOM relationships. The following steps are carried out in the configuration process:

1. determine the generic product the customer wants to configure; this generic product may be a finished product, a subassembly or a component;
2. assign values to all of the parameters associated with this generic product;
3. check that all of the constraints associated with this generic product are satisfied;
4. execute the two previous steps for all of the lower levels of this generic product, thereby passing all relevant parameter values to these lower levels;

The specification is complete when the complete structure has been traversed in this way. The result of the configuration process is a list of parameter values for all of the generic products in this structure. This list specifies the chosen product variant uniquely from a customer's point of view.

The generator uses these parameter values to produce a customer specific bill of material from the product family structure. Provided that values are assigned to all of the parameters associated with the generic primary products (GPPs) and no constraints are violated in the configuration process, all of the necessary primary variants can then be determined. When the required variants of generic secondary products (GSPs) are available in stock (e.g., when the customer order decoupling point is not at the level of the primary variants), they are allocated for use in completing the customer order. If these secondary product variants are not available in stock, however, then they are manufactured specifically for this customer order. Figure 26.4 shows where the customer order decoupling point could lie in a product family structure which consists of GSPs and GPPs.

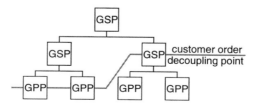

Fig. 26.4 Customer order decoupling point

In the next sections, we will explain in more detail how product family data is compiled during the product development phase. We will demonstrate how the generic bill-of-material approach can be used to structure the product family data as it is accumulated. Parameter values can be regarded as product specifications in this sense; the product family structure and the primary product variants can be used as the basis for planning and monitoring the realisation of these specifications. Furthermore, better design decisions can be made based upon improved insights into the different points of view for a product family. For this reason we have placed special emphasis on the design rules which take the complexity of development and logistic issues into account.

26.4 DEVELOPING PRODUCT FAMILIES

As mentioned in Chapter 10, product families have been defined in order to be able to address a wide variety of customer requirements using only a limited set of physical modules. This objective implies that the definition of a product family needs to take the market platform as well as the technical product architecture into account. These aspects are discussed in this section.

The term "product family" can be confusing since many companies consider any range of similar products to be a product family. A closer examination of such a range of products may reveal that a definition of a product family based upon product similarities provides little or no benefit or added value. There is an overhead cost to develop and maintain a range of products, regardless of whether the products are related or unrelated. This cost often exceeds the benefits when product families are created through an evolutionary product development process or when insufficient consideration is given to the aspects of the product life cycle [Onkvisit, 1989].

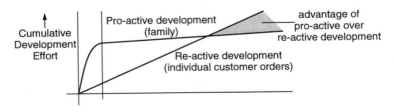

Fig. 26.5 Proactive versus reactive development

Figure 26.5 shows the typical difference in development effort between proactive and reactive development approaches. A reactive development approach is where new product variants are defined as required in response to individual customer orders. This often leads to an undesirable situation at the end of the product life cycle when newer product variants have very few aspects in common with the initially developed variants. The cumulative development effort increases, nevertheless.

In contrast, a proactive development approach is based upon the initial formulation of a life cycle for a product family and a subsequent development of the product variants within a relatively short time frame. The architecture of such a product family is determined in such a way that the product family and the cumulative development effort will remain stable for a certain period of time. This approach generally requires the existence of a relatively mature product/market combination in which all parties involved have a good understanding of the "market platform" and the architecture of the new product family.

For mature product/market combinations, a number of design rules can be formulated to reduce the complexity of the development process and the operational logistic process. These design rules can be summarised as follows and are explained in more detail in the next section:

- State the specifications clearly in terms of parameters, parameter values and constraints which have been agreed by all of the involved parties such as marketing, manufacturing, logistics and service.
- Design a stable architecture for the product family, including possibilities for integrating future modules with the existing modules; this should take the manufacturing and service requirements into account and make allowances for the life cycle considerations of all major components and modules.
- Provide a comprehensive mapping of the parameter values to the physical architecture in order to enhance the modularity of the product family. A highly modular product family reduces the complexity of development and manufacturing activities, facilitates engineering changes and makes servicing easier [Ulrich and Tung, 1991].

26.5 DESIGN RULES

For the purpose of this section, we assume that the development process is supported by an engineering database for integrated product development. The data model presented in Section 26.2 provides the basis for storing and retrieving the generic bill-of-material data used by the different points of view [Erens and Hegge, 1993]. Furthermore, this data model is used to clarify the different types of design rules.

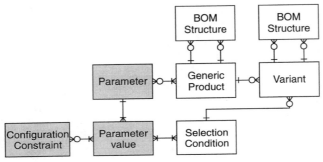

Fig. 26.6 Defining the market platform

Market platform

An initial definition of the market platform can be formulated in terms of parameters, parameter values and constraints.

In information technology terminology, the market platform is defined by populating the data entities which represent the *parameters*, *parameter values* and *configuration constraints*. These entities can be regarded as specifications for the development of a product family. The relationship between the *parameter* and *generic product* entities in the data model implies that the product family specifications cannot be finalised without taking the definition of the problem solution (i.e., the generic products and their structure) into account. The development of a car can be used as an example. In this case, a parameter called ABS (with a value of either "yes" or "no") would normally affect several generic products within the family structure. On the other hand, the technical realisation of ABS will normally be highly dependent upon the architecture chosen for the car.

The complexity of the product development activities can generally be reduced by limiting the number of specification variables. In other words, the number of parameters, parameter values and constraints on parameter values should be restricted. Although the number of constraints can often be reduced by improving the modularity of the design, the number of parameters and the range of parameter values are more difficult to reduce since these are dictated primarily by the market.

Nevertheless, manufacturers have a tendency to increase the number of product variants because of their uncertainty about customer requirements. The best strategy in this case may be to postpone the development of product variants until customer requirements have been verified. Furthermore, the architecture of the product family should be structured in such a way that the definitions of such "volatile" specifications can be easily incorporated in the architecture. Examples of volatile specifications can be found in consumer electronics products where special "features" and "styles" are dependent upon current trends and fads.

Architecture

In many cases, the architecture of a product family is designed in such a way that it can be reused by a new generation of products. Cars, for example, are generally considered to be mature products in which modules and the structures of these modules are retained in new generations. The generic product, BOM structure and variant entities thus become extremely useful for project management purposes. Historic information on development lead times, module interfaces, costs and problems encountered in the past can all be stored and maintained based upon a data model structure as shown in Figure 26.7.

The architectural complexity of a product can be reduced by limiting the number of data entities and their relationships. This can be

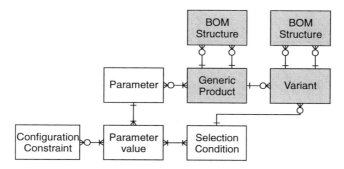

Fig.26.7 Design of the physical architecture

accomplished by limiting the number of generic products, simplifying the structure (especially the depth) of the product family and reducing the number of primary variants which would otherwise lead to the definition of an excessive number of product variants.

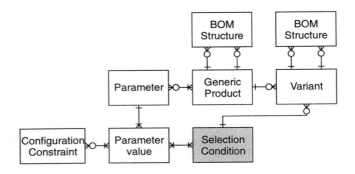

Fig. 26.8 Mapping of parameters to the physical architecture

Mapping

The entity in the data model which reflects the complexity of the development process to the greatest extent is the *selection condition*. This entity provides a formal description of the mapping of parameter values to primary variants. In other words, this entity translates the specifications into a physical product.

A selection condition for a primary variant is formulated in terms of parameter values and Boolean operators such as AND, OR, NOT and ⇒. Selection conditions provide a useful basis for engineering change control in addition to their usefulness as specifications for the development of a product. The consequences of changes to a primary variant with respect to the market platform (i.e., from the customer's point of view) can be identified accurately and easily in terms of changes to the selection conditions.

From the "crow's feet" in the data model, it can be seen that there are four possible relationships between parameter values and primary variants:

- where a parameter value (specification) is used in exactly one primary variant (physical product). This 1:1 relationship reduces the complexity of development considerably. This can be seen as a modular design since a large variety of products can be configured

from a limited set of modules. This approach is always useful with respect to parameter values, but not necessarily for specifications in general. If every specification option is associated with a separate physical module, the cost of the product will be higher than in a situation where many specifications are integrated into a multi-functional module.
- where multiple parameter values influence precisely one primary variant. This N:1 relationship has two major drawbacks. The first drawback is the large number of primary variants required. Suppose, for example, that three parameters are associated with a given GPP, and each parameter has two possible values. This generic product is likely to have 2x2x2=8 primary variants unless constraints prohibit certain combinations of parameter values.

The second drawback concerns the time which may be needed to complete the development phase. If we assume that each parameter value represents a different product specification, a product variant can only be developed after all of the relevant parameter values have been assigned. If one of the parameter values remains unspecified, then the development of a product variant is likely to be delayed.

In the case of product variants in the form of software options, integration using N:1 relationships is attractive since the cost of software is determined largely by the development costs. With this in mind, a single development effort could result in producing a product in which all of the software options are pre-installed, but locked and made unavailable to the customer until payment is received for a certain software configuration. The customer then receives the code to unlock the specific software configuration which was ordered. In this way, software can provide a large number of product variants based upon only one standard technical variant.

- where a single parameter value is associated with multiple primary variants. An example of such a 1:N relationship could be installing an ABS feature in a car whereby several physical modules are required. The complexity of such a function stems from the integration of the function in the overall product [Ulrich and Tung, 1990]. If it is assumed that the necessary modules needed to implement the ABS feature are distributed over various sections of the car, then a poorly executed integration function would threaten the integrity of the overall product.

- where multiple parameter values are associated with multiple primary variants. This N:M relationship is the worst type of situation since it incorporates all of the drawbacks of the 1:N and N:1 relationships. A product which possesses many N:M relationships can be regarded as being extremely non-modular. This would make it difficult to isolate functions within the physical product since the module interfaces would not correspond to functional boundaries.

Three simple design rules can be formulated:

- when possible, each parameter value should be used in only one primary variant and the other specification aspects should be integrated within multi-functional modules;
- specifications which are dependent upon current trends and fads and which can only be defined close to the product family introduction date should be used in such a way that they can be incorporated in the physical architecture at a later stage;
- from a manufacturing point of view, all primary variants should be assembled relatively late in the production process. This permits the production of a common product until the final stages of manufacturing or, in the case of software, through to and including product installation.

Modularity

We can measure the modularity of a product from the relationship between the market platform and the physical architecture; see also [Erlandsson, 1992] and [Erixon, 1993]. A product family is fully modular from a product variant point of view if all primary variants can be combined in an arbitrary way to create finished product variants.

In a typical situation we could assume that the multiplication of primary variants theoretically gives M finished product variants, but that there are actually only N commercially available finished product variants due to constraints on combinations of parameter values and due to the fact that some parameter values are used in two or more primary variants. The following formula can then be used as a measure of the modularity of a product family:

$$\text{Modularity} = \frac{N}{M} = \frac{\text{Actual number of variants}}{\text{Theoretical number of variants}}$$

This formula could be used in the product development phase to measure an increase or decrease in performance. A value of "one" or more (in the case of software) is ideal.

26.6 CONCLUSIONS

A possible implementation of the generic bill-of-material approach, translated into information technology terminology, was presented in this chapter. Special emphasis was placed on the structure of the data which is used by generic bill-of-material applications. The data model used here also served as a basis for formulating design rules for the development of a product family.

We can conclude that a generic bill-of-material and its formalisation in the form of a data model provide insights into the dependencies between the market platform (specifications) and the physical architecture (physical product). By following certain design rules, the number of dependencies can be reduced to improve the quality of the product and to simplify product development activities.

REFERENCES

1. Griethuysen, J.J. van, *Concepts and Terminology for the Conceptual Schema and the Information Base*, ISO/TC97/SC5 N695,1982.
2. Hoekstra Sj., Romme J.A.C., *Internal Logistic Structures*, McGraw-Hill, 1992.
3. Hegge, H.M.H. and J.C. Wortmann, Generic bill-of-material: a new product model, *International Journal of Production Economics* 23, pp.117-128, 1991.
4. Hegge, H.M.H., A Generic Bill-of-Material Using Indirect Identification of Products, *Production Planning and Control*, vol. 3 no. 3, pp.336-342, 1992.
5. Onkvisit, S. and J.J. Shaw, *Product Life Cycles and Product Management*, Quorum Books, 1975.
6. Ulrich, K. and K. Tung, Fundamentals of Product Modularity, DE-Vol. 39, *Issues in Design Manufacture Integration*, ASME 1991.
7. Erens, F.J., H.M.H. Hegge, E.A. van Veen and J.C. Wortmann, Generative bills-of-material: an Overview, *Integration in Production Management Systems*, H.J. Pels and J.C. Wortmann (eds), Elsevier Science Publishers, IFIP, 1992.

8. Erens, F.J. and H.M.H. Hegge, Manufacturing and Sales Co-ordination for Product Variety, *International Journal of Production Economics*, 1993.
9. Erixon, G., B. Östgren and A. Arnström, Evaluation Tool for Modular Design, *The International Forum on Product Design for Manufacture and Assembly*, June 1993.
10. Erlandsson, A., G. Erixon and B. Östgren, Product model - the link between QFD and DFA?, *The International Forum on Product Design for Manufacture and Assembly*, June 1993.

27

Standard software packages for business information systems in customer driven manufacturing

27.1 AN ARCHITECTURE FOR BUSINESS INFORMATION SYSTEMS

This chapter will discuss business information systems for customer driven manufacturing. It should be recalled from Chapter 6 that there are many types of customer driven manufacturing. Depending upon the type, customer driven activities are combined with manufacturing of standard products. Almost each type of customer driven manufacturing requires some form of inventory control for standard products as well. Furthermore, actual practice shows companies seldom fit nicely in a typology. Rather, companies have different lines of business which results in a kind of mixture when their business is to be characterized.

In order to simplify the discussion on business information systems, we will concentrate on the difference between the product-oriented make-to-stock situation (in section 27.2) and the engineer-to-order situation in general (in section 27.3). For both situations, we will present information systems architectures and elementary data structures. The data structures will be used to highlight differences. Conclusions will be presented in section 27.4.

This chapter should be considered as an introduction. In subsequent chapters we will discuss the nature of customer-driven engineering and production in more depth. More specifically, Chapter 28 will discuss *document management* in customer driven manufacturing. Chapter 29 focusses on *engineering as a process to be controlled* in case of product-oriented engineer-to-order manufacturing. Chapter 30 discusses shop floor control information systems.

In order to explain the nature of business information systems for manufacturing in general, we will make use of an architecture for business information systems which has been proposed by Bertrand et al. [1990] (Chapter 6). This architecture is presented in figure 27.1. This figure shows that a business information system can be viewed as a number of concentric circles representing various system layers.

The innermost circle represents the systems platform. It supports different system programs such as an operating system, a data base management system and a query language. The subsequent layers represent the different user applications.

The applications which support the registration of *state-independent* data are found on the second layer. This data describes the products, technology, resource capacities, operations, etc. which are independent of the state of the transformation process at any particular point in time. Examples of state-independent data include: standard products, bills of materials, standard routings, resource capacities, standard throughput times, etc. State-independent data can refer to *product* data as well as data. State-independent data provide a basis for the registration of state-dependent data.

The applications for the registration of *state-dependent* data are found on the third layer which rests on top of the state-independent data. These data describe the state or condition of materials and orders in the transformation process which need to be controlled. Materials are received, inspected, stored, issued, consumed for assembly and so on. Other resources, such as equipment, tooling, or fixtures are characterized by time-phased availability. Customer orders start as prospects, and are transformed into confirmed orders, completely specified orders, shipped orders, invoiced orders and finished orders. Internal work orders and orders issued to suppliers and subcontractors have a similar life-cycle. This layer does also contain software to store and retrieve forecasting and planning numbers.

Finally, the outermost layer is comprised of the applications which make use of the data registration systems found in the other layers. To start with, *decision support* systems are found here which provide support to the human decision process. In connection with production control, emphasis is usually put on decision support systems. Next, structured decision systems which are able to carry out a decision process without human intervention are found in this layer. Finally, this layer contains workflow applications which may guide users through a repeating pattern of activities required to execute a business process. Examples of such business processes are the release of a work order, the shipment of a customer order, or the payment of an invoice.

In addition to the different layers which have been defined in this way, the information systems' architecture also provides a structure for distinguishing between various control levels. The upper half of each layer is reserved for the applications which support the goods flow control (GFC) level. The lower half of each layer is used for the applications which support the production unit control (PUC) level. In this way it should be clear that the data registration systems (and the data which is registered) as well as the decision support systems can be different at each control level.

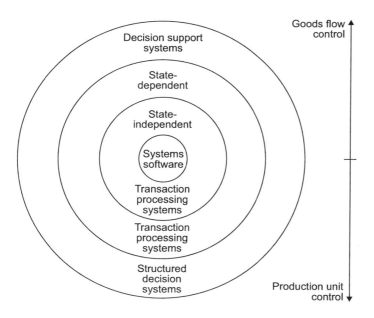

Fig. 27.1 A general architecture for business information systems

27.2 INFORMATION SYSTEMS FOR PRODUCT-ORIENTED MAKE-TO-STOCK PRODUCTION

27.2.1 General remarks

For reasons of simplicity, we will base the discussion of information systems for make-to-stock production on standard software packages for Manufacturing Resources Planning (MRP II). This can be criticized of course, because there are several other approaches. However, the conclusions which we draw for data structures remain valid for several other approaches, such as OPT, Period Batch Control etc. MRP II is nowadays often combined with Distribution Resources Planning (DRP II). When logistics in a network of warehouses and factories is supported by a software package, the term Enterprise Resources Planning (ERP) is becoming popular. ERP packages are supposed to have closely interlinked modules available for financial control, quality control, human resources, commercial activities and other business processes. In this chapter, our focus will be on manufacturing.

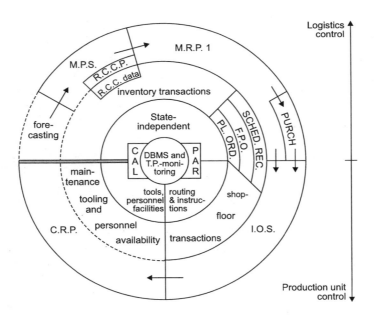

Fig. 27.2 Architecture of MRP II systems

Information systems for product-oriented make-to-stock production 337

The architecture of MRP II standard software packages is depicted in figure 27.2. Note that this figure is split into two areas: the upper part represents the Goodsflow Control (GFC) area, whereas the lower part represents the production unit control (PUC) area.
The implementation process of a standard software package requires that a considerable number of parameters be initialised. This is shown in figure 27.2 (abbreviated as PAR). Examples of such parameters are e.g. default values for many attributes, unit-of-measure convention, scheduling parameters, lot-sizing parameters, parameters for handling exception conditions, etc. Another typical set of data which is often overlooked, is the calender file (CAL).

27.2.2 State-independent transaction processing systems

In MRP II-software packages, state-independent transaction processing systems first of all deal with product information. Usually, companies have a file at disposal in which all standard items are represented by one record. State-independent attributes of such items are numerous, e.g.:

- identification number and description
- classification codes for value, commodity type, geometry, material, primary storage location, unit of measure, ordering policy, safety stock, safety lead time, stocking policy, packaging instructions, etc.
- standard leadtime (SLT)
- references to prime vendor, prime buyer, prime responsible planner
- various cost attributes.

Often, it is necessary that these attributes have received a meaningful value. For example, if a purchasing module is operational, many systems require that all items should be classified either as a purchased item or as a non-purchased item. In the case of a purchasing item, the prime purchaser should be known and the commodity code, and so on. Of course, certain software packages would provide default values for mandatory attributes, but the general experience with MRP packages is that full usage of all possibilities in available applications will require quite some data-entry effort when new products are defined. Finally, a point to bear in mind is that all items defined in MRP II-packages are standard items.

Bills of material (BOM)

The BOM-concept was already introduced in Chapter 5 (cf. figure 5.5a). The previous chapter (Chapter 26) was entirely devoted to bills of material. A more detailed description of standard functionality and standard data structures is given here, as an introduction for readers who are not familiar with the details.

A bill of material for a parent item is a list of components required for this parent item. More specifically, a bill of material represents a set of parent-component relationships, where each such relationship is an entity of itself with at least the following attributes:

- parent identification number (key attribute)
- component identification number (key attribute)
- effectivity dates
- quantity of component per unit of parent
- yield/scrap factor
- lead time adjustment.

Several short comments are appropriate here to give the reader some impression of the bill-of-material processor ususally encountered in MRP II packages.

Firstly, a component item which occurs in a bill of material of some parent may also act as a parent itself. In this way a multi-level bill-of-material structure may be defined. However, an item is never allowed indirectly to become a component of itself. In other words, if the multi-level bill-of-material structure is seen as a directed graph, this graph should be cycle-free. The bill-of-material processor should check this fact at each change in parent-component relationship. Secondly, a component item may occur as a component of several parents.

Thirdly, the effectivity-date attributes raise an important topic in state-independent transaction processing, viz. the fact that product structures suffer from so-called engineering changes. An engineering change represents a change in the way a product is designed, manufactured or ordered. Some systems support the flow of engineering changes by including an entity type ENGINEERING CHANGE in the database. We will ignore this issue in the remainder of this chapter. However, the next chapter (Chapter 28) will pay due attention to engineering changes.

Routings

The concept of a routing was already introduced in Chapter 5 (cf. figure 5.5b). The routing of an item is an ordered list of normative operations required for the manufacturing of this item from its components. A normative operation consists of a number of manufacturing steps, but these steps and their description is usually not formally represented in the database that supports an MRP package. Therefore, a normative operation is an entity that has at least the following attributes:

- item identification number (key attribute)
- routing sequence number (key attribute)
- capacity unit: this denotes the (machine) capacity where the operation should take place
- normative amount of capacity required for the operation (sometimes called operation time). This amount is usually computed from more detailed (normative) attributes, such as set-up time, run time per piece, time required to fetch and inspect tooling etc.
- transportation time allowance. This refers to transportation of a batch from one operation to another.
- waiting time allowance. This attribute is used in scheduling the operations of a routing in order to take into account queueing of a batch at a capacity unit.

From these attributes, the scheduling routine for shop floor control in the production unit is able to compute a lead time for a work order for a specific item with a given lot size. This lead time is the sum of transportation time allowances, waiting time allowances and computed operation times, for all routing operations for the item in question. This computed lead time should of course be equal to the standard lead time (SLT) attribute of the items (however, only a few MRP II-packages maintain this equality).

A routing is a precise description of the flow of a work order through the factory. It is used to evaluate the impact of planned orders generated by MRP I software on detailed capacity load profiles.

27.2.3 State-independent data structure diagram for MRP

In order to make the above discussion more specific from an information systems point of view, we shall concentrate on data structures. In

Bertrand and Wortmann [1981] it is argued that data structures consitute the heart of an information system. The pictorial notation used here is derived from Martin [1986].

The data structure diagram for the state-independent part of MRP packages (in their most elementary form) is given in figure 27.3. First, consider the entity type ITEM. The figure shows this entity type at the top right. This figure shows that the manufacturing of an item requires a number of operations and a number of components.

The N-to-1 relationship type from NORMATIVE OPERATION to ITEM represents the routing of an item. Furthermore, it specifies that each NORMATIVE OPERATION is always related to exactly one item. The N-to-1 relationship type from NORMATIVE OPERATION to CAPACITY UNIT specifies that each normative operation should always be related to precisely one capacity unit.

The entity type GOZINTO RELATION represents parent-component ("goes into") relations in the bill-of-material structure. Each "goes-into relationhip" is associated with two items: one parent item and one component item. Therefore, two relationship types have to be defined from the entity type GOZINTO RELATION to the entity type ITEM. These relationship types are called "explosion" and "implosion". Via

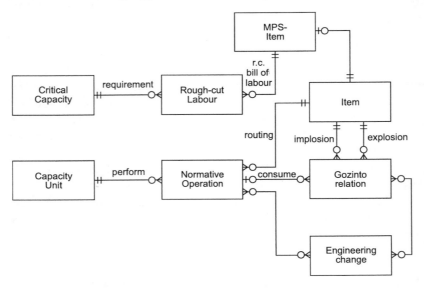

Fig. 27.3 State-independent data structure diagram for MRP software packages

the "explosion" relationship type, each (parent) item is related to zero, one or more incoming "goes-into-relations". Via the "implosion" relationship type, each (component) item is related to zero, one or more outgoing "goes-into relations".

A software module that enables the end user to maintain the information described in figure 27.3 is called a bill-of-material processor. This type of module provides the end user with a number of screens to perform queries and updates, but it also enforces adherence to a number of constraints. One of these constraints states that the bill-of-material structure should be cycle-free.

27.2.4 State-dependent transaction processing systems.

The third layer in figures 27.1 and 27.2 is concerned with state-dependent transaction processing systems. In production/inventory control, this means the recording of orders, forecasts and budgets, and time-phased availablity of resources (materials and capacities).

In MRP packages, orders can take the form of customer orders, MPS orders, (firm) planned orders for MRP items, and released orders. Materials can be on hand in inventories, they can be in transit, or on the shop floor, as floor stock or work in progress. If material is being consumed, it is usually for released orders. The flow of orders and the flow of materials meet each other in the released orders.

In this subsection we shall first describe the flow of materials, and then the flow of orders.

Within classical MRP packages, the flow of material is always recorded in terms of items (part numbers). An item uniquely identifies materials, and two material occurrences with the same item-number are exchangeable from a logistics point of view.

The fact that all physical materials are identified in MRP systems by part numbers has important consequences. For example, it is difficult to represent the situation where different parent items result from one work order in MRP systems. Such a situation occurs in Group Technology, where different parts from a common family are processed together in one batch. A similar situation occurs in industries, where sorting different products from the same lot leads to several parent items being produced in one work order. Yet another example is presented by "by products" which often inevitably appear in process industries. None of these situations are easily treated in classical MRP

II software packages. They require (considerable) extensions, which are briefly discussed in Wight and Landvater [1983].

A customer order is a list of standard items in MRP systems. For each standard item, an order quantity is specified. A due date, price, and other delivery conditions may be added. However, the reader should note the difference with customer orders in customer driven manufacturing, to be described in the next Section.

27.2.5 State-dependent data structure diagram for MRP

The reader may refer to the data-structure diagram of figure 27.4 as an aid for the subsequent discussion. This Figure is an extension of figure 27.3. According to MRP theory work orders are either planned orders, firm planned orders, or scheduled receipts. (Some systems do distuinguish several more types but these three are sufficient for our discussion). The three types of work orders are shown together at the right of figure 27.4.

For ease of representation, figure 27.4 treats planned orders, firm planned orders and scheduled receipts as being one entity type (work order). Generally speaking, this is incorrect. Planned orders do not

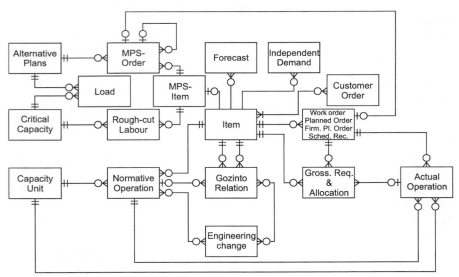

Fig. 27.4 State-dependent data structure for MRP II

have a specific identity or a unique identifier. They can be changed by MRP at any time.

Firm planned orders are quite different. They are identified by a key which is known to the user. They can be changed in many respects: lot size, lead time, exploded gross requirements, due date, and so on. Scheduled receipts, finally, are different again from firm planned orders. Exploded material requirements for scheduled receipts are initially posted as allocations. These allocations gradually disappear when component materials are issued. Value added for scheduled receipts is maintained when material issues and operation completions are reported.

All three types of work order have a routing of actual operations, as shown in figure 27.4. For planned orders, this routing is a copy of the normative routing of the item concerned. For firm planned orders, the routing may be altered. After order releases, the routing can no longer be changed any more in most cases (although some systems do allow operations which differ from the actual routing to be reported ready for a work order).

In the above discussion, work orders are implicitly considered to be internal work orders. However, similar comments can be made for purchasing orders and subcontracting orders.

Work orders are a specification of time-phased supply. For each item, the purpose of MRP is to balance time-phased supply with time-phased requirements. Therefore, we shall briefly discuss the way in which time-phased requirements are specified. Gross requirements are either dependent on or independent of demand. Dependent demand is generated by explosion from work orders, as we have seen. The reverse information, specifying the source of demand for a particular allocation or gross requirement, is called pegging.

Independent demand, on the other hand, may take the form of customer orders or forecasts. For the sake of simplicity, the data structure for representing independent demand is not included in figure 27.4.

27.3 INFORMATION SYSTEMS FOR ENGINEER-TO-ORDER PRODUCTION

27.3.1 General remarks

Various types of engineer-to-order production have been introduced in Chapter 6. Both the producer of packaging machines of Chapter 9 and the shipbuilding industry of Chapter 12 provide case studies of engineer-to-order companies. Chapter 21 has elaborated on production control for the engineer-to-order case described in Chapter 9. In the present chapter, we will discuss information systems for this type of engineer-to-order production.

Recall that Chapter 22 continued the case described in Chapter 21, but now for the customer-driven engineering part only. Accordingly, Chapter 29 will proceed with IT-support for the management of engineering in the case described in Chapter 22, viz. packaging machine manufacturing (which is product-oriented engineer-to-order manufacturing).

Engineer-to-order manufacturing differs from the make-to-stock manufacturing. In the make-to-stock situation, the products and routings are given, but the precise timing and quantity of demand is uncertain. In the engineer-to-order situation, the timing and quantity of demand are given, but the precise nature of products and routings are uncertain.

In many order-driven production systems, the supply of materials is not a critical issue. For example in the construction industry, many materials can be delivered at short notice, whereas major equipment and human resources need to be planned further in advance. Consequently, it is unnatural to derive activities from bill-of-material structures and quite natural to do the reverse: derive material planning from activity planning.

The general architecture with three distinct application layers, as illustrated in Figure 27.1, can be used to develop an information architecture for customer order driven engineering. First, a further explanation of the state-independent data and the state-dependent data is needed. Subsequently, decision support will be described.

Information systems for engineer-to-order production 345

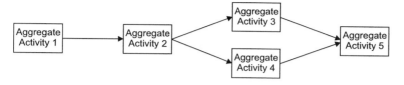

Fig. 27.5a Reference project as a network of reference aggregate activities

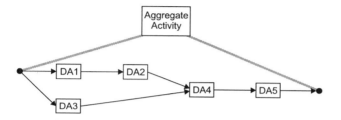

Fig. 27.5b Aggregate activities as a network of detailed activities

27.3.2 State-independent transaction processing systems

State-independent data in engineer-to-order production differ from state-independent data in production of standard products. State-independent data in customer order driven engineering can be characterized as being primarily *reference* data, i.e. data which is searched and retrieved so that it can be modified or extended for use in a specific application. In general, these data cannot be used directly by an application since these data must first be accessed and interpreted by a specialist. It is necessary to explain exactly what is included in the registration of state-independent data in connection with engineer-to-order production before the nature of the state-independent data can be discussed.

In the first place, this data category includes data objects which describe *products*. (Of course, this data category is most important in product-oriented companies). All of the accumulated knowledge concerning product descriptions is registered here so that it can be referenced easily for use in developing the specifications for customized products. These "reference" product descriptions can be seen as the results of previous customer-independent engineering activities which are intended to be used to support the engineering of customized products. Included here are not only reference drawings and bills of materials, but also descriptions of product structures in the earlier stages of development (e.g., functional structures, see Chapter 29).

In the second place, registration of state-independent data includes *workflow* data. (Of course, this data category is most important in workflow-oriented companies). The data which are maintained here concern the customer driven engineering process as well as the subsequent physical production process.

These data are reference projects, which consist of reference aggregate activities (see fig. 27.5a). Each reference aggregate activity may be decomposed into reference detailed activities (see figure 27.5b). This principle may be repeated, of course, but such repetition does not change the principle. (However, the shipyard of Chapter 12 had even five levels of aggregation!)

These state-independent workflow data comprise data which are known to be the same for each occurrence of the process. By maintaining the data in this way, it is unnecessary to re-invent the information repeatedly for each new customer order.

Thirdly, state-independent description of *resources* data-objects should complete the picture. (Of course, this data is most important in resource-oriented companies such as the shipyard of Chapter 12).

Already in Chapter 5, a distinction was made between capacities and capabilities. This distinction can be illustrated by the example of rough planning of the shipyard in Chapter 12, which is based on questions such as whether or not a heavy crane or a large hall is available.

Chapter 17 has given a detailed description of (human) resources in customer driven systems. Broadly speaking, resources in customer driven systems tend to be less specialised than resources in make-to-stock production environments. Resource requirements are considered to be requirements for certain *skills* rather than requirements for certain *persons*, and requirements for *capabilities* of machines and tools rather than for certain designated machines and tools.

A distinction can be made between the objects which are required at the GFC level and the objects required at the PUC level. This distinction pertains to the product descriptions as well as to the workflow data objects and the resources data objects. The required state-independent data are outlined in general in Table 27.1.

	Workflow data	Product data	Resources data
GFC-level	reference network of aggregate tasks	reference product descriptions reference GFC-products structures	PU-structure critical capacities and capabilities
PUC-level	reference activities of routings	reference product structures for testing, assembly or installation	detailed description of capacities and capabilities (skills)

Table 27.1 State-independent objects for customer driven manufacturing in general

	Workflow data	Product data	Resources data
GFC-level	reference network of aggregate (engineering) tasks	reference descriptions of engineering deliverables	PU-structure within engineering critical engineering capacities
PUC-level	detailed description of engineering tasks through reference activity network	reference descriptions of engineering documents	detailed description of engineering capacities and capabilities

Table 27.2 State-independent objects for customer driven engineering

The interesting point in engineer-to-order companies lies in the fact that the objects of Table 27.1 should be stored not only for physical production activities but also for non-physical activities: the engineering work.

The state-independent data objects required for engineering are presented in Table 27.2. The state-independent data objects which are required for subsequent physical production are listed in Table 27.3. Aggregate reference data are required at the GFC level to specify how the customer driven engineering process and the physical production process should be carried out. For this reason, the PU structure, the critical resource capacities and standard networks of aggregate activities are documented for both processes. In addition, information is required at this level about reference engineering documentation and reference models for product structures. Reference engineering documents provide examples for the separate GFC items (e.g., a preliminary draft of the design) which are used to control the process at the GFC level (see also Chapters 22 and 29).

Reference product structures are needed for determining, for example, throughput times and quotation prices. Product data for the physical production process can also be derived from these reference product structures (see Table 27.3).

	Workflow data	Product data	Resources data
GFC-level	reference network of aggregate (production) tasks	reference product descriptions reference GFC-product structures	PU-structure critical capacities
PUC-level	reference production routings	reference product structures for testing, assembly or installation	detailed description of capacities and capabilities in production

Table 27.3 State-independent objects for customer driven physical production activities

27.3.3 Data structure diagram

We shall now discuss the state-independent data structure for a product-oriented engineer-to-order company in more detail. Our discussion is visualised in the data structure diagram of figure 27.6, which is comparable to figure 27.3.

Let us start the discussion with the description of *products*. Consider the entity type PRODUCT FAMILY, at the top of Fig 27.6. This entity type contains a few descriptive attributes for each product family which the company offers to the market. A product family represents systems that are composed of a number of main components (cf. Chapters 9, 11, 12). Different realisations of these main components in different customer orders will be quite different. But the common part can be described in an entity of type REFERENCE FINAL PRODUCT, with an

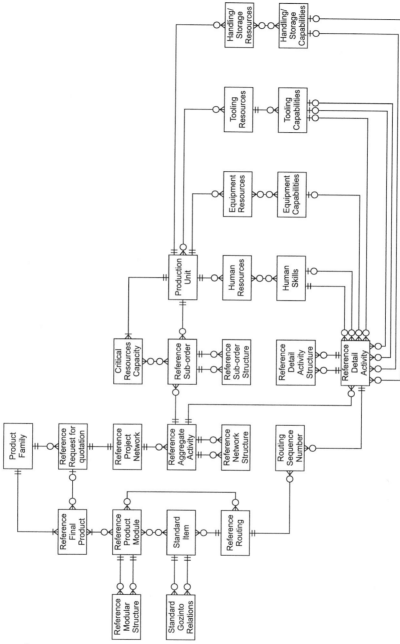

Fig. 27.6 State-independent data structure diagram for resource-oriented engineer-to-order production

underlying product-structure described in REFERENCE PRODUCT MODULE and REFERENCE MODULAR STRUCTURE. Somewhere in the bill-of-material, standard items may be used. This is covered by the entity types STANDARD ITEM and STANDARD GOZINTO RELATION, just as in MRP. Standard items have one routing, whereas product modules may be linked to several routings. This illustrates the fact that reference information is interpreted and used by humans before being actualized for customer orders.

In the previous dicussion, generic bills of materials could have been included, as decribed in the previous chapter. For reasons of simplicity, this has been omitted. The description of engineering deliverables (see Table 27.2) is deferred until Chapter 29.

Let us now consider the *workflows*. First of all, figure 27.6 shows a new type entity type REFERENCE AGGREGATE ACTIVITY, connected with the entity type REFERENCE DETAIL ACTIVITY. A reference aggregate activity consists of a network of reference detail activities. It is not necessary for each reference activity to be completely specified. A reference aggregate activity with its reference detail activities is a kind of provisional structure, which may be used in several projects to characterise actual work which has not yet been specified in detail. An actual aggregate activity within a project may be related to zero or one reference aggregate activities. Definition of a specific aggregate activity structure may be deferred until the time the actual work for a specific customer order is started.

To conclude our discussion on workflows (i.e. routings and activities), it can be summarised as follows. The information system outlined here should be considered primarily as a support tool for engineers who plan future work. It may be helpful for a project engineer who has to define a specific network of aggregate and detail activities or for a manufacturing engineer who is involved in work preparation for shop floor operations. The support of the manufacturing engineer is given in a manner that provides both flexibility and standardisation. There is a sharp distinction with MRP packages, which require fully specified bills of material and routings, because they have to generate automatically planned orders and capacity load profiles in MRP I and CRP programs.

Finally, a discussion on *resources* is worthwhile. First of all, figure 27.6 distinguishes between human resources, equipment resources, tooling resources, and resources for material handling and storage. This reflects the discussion of Chapter 17, which stressed the specific nature of human resources, but also the discussion of Chapter 12, which showed the necessity to have material handling and storage available in addition to certain machines. In the previous sub-section, *human skills* were considered to be essential properties of humans. These skills are explicitly modelled in figure 27.6. The structure allows to be represented the fact that one or two skills are required by a reference activity. Many humans may have those skills, as indicated by the N:M relationship between HUMAN RESOURCES and HUMAN SKILLS in figure 27.6.

For the sake of comparison, similar structures have been chosen for the other resource types and their capabilities. A reference activity may require one piece of equipment, up to three tooling capabilities and up to two handling/storage capabilities in figure 27.6.

The real work orders which will eventually accompany the material transformations on the shop floor will gradually become known. The material requirements are also at best partially known in the beginning: they become more pronounced during the product-engineering phase. The work orders (i.e. the detailed activities of a manufacturing task) will gradually become known during the manufacturing engineering stage, when the work preparation is performed. These work orders, again, are not necessarily related to an item in the item file (not even to a generic item).

27.3.4 State-dependent transaction processing systems.

There are several issues to be discussed in connection with state-dependent transaction processing:

- the importance of the customer order
- the fact that actual workflow and product data become known during the project
- the existence of work orders without having associated items.

These issues will be covered below. However, for the clarity of discusion, we will not elaborate here on product descriptions or resource descriptions. Generic product descriptions have been discussed in Chapter 26, and state-independent product descriptions will be treated in more length in Chapter 29. State-independent resource descriptions have been covered above, and the essentials are not different in the case of state-dependent data.

Customer-orders

In engineer-to-order information systems, *quotations* and *customer orders* play an important role. The problem of keeping records of past customer orders (so-called configuration management) is interesting and important, but it would distract the attention from the main points to be made here. We will therefore not consider the issue here, but focus on customer orders for new projects.

The definitions of customer orders and quotations in an engineer-to-order environment highlight some of the differences with standard products manufacturing. In classical MRP packages, quotations are not known and customer orders are lists of known items to be delivered in specific quantities at a specific date. In an engineer-to-order company, a quotation is a complex volume of requirements and solutions, with drawings and computations included. A customer order starts as a copy of a reference network, and gradually becomes a complete set of product documentation, process documentation and resource requirements specification. The next paragraphs will further elaborate this point.

Gradually emerging data

If a customer order is entered in the system, it is usually related to a reference network. When the customer order is firmly accepted, a copy of this reference network is created and pegged to the customer order. This is shown in the data structure diagram of figure 27.7, which is an extension of figure 27.6. The real work orders which will eventually accompany the material transformations on the shop floor will emerge gradually. Material requirements are also at best partially known in the beginning; they become more pronounced during the product-engineering phase. The work orders (i.e. the detailed activities of manufacturing) will gradually become known during the manufacturing engineering stage, when work preparation is performed. These work orders are again not necessarily related to an item in the item file (not even to a generic item).

If the blank attributes of the activities in the customer order network have been specified then earliest start dates, latest start dates, etc., can be computed.

The activities usually pass through the following stages:

- not yet completely specified (unsuitable for normal network planning techniques)

Information systems for engineer-to-order production 353

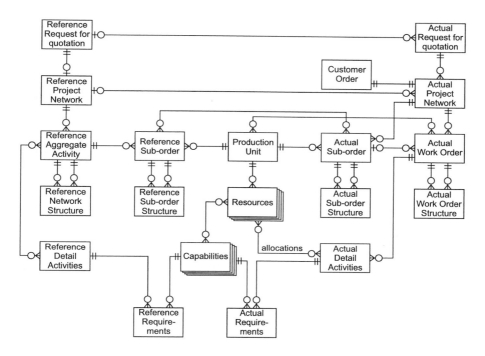

Fig. 27.7 State-dependent data structure diagram for resource-oriented engineer-to-order production

- completely specified
- firm planned (i.e. it has a specific start date and due date)
- partly ready
- ready
- deleted.

Work-orders without having associated items

In MRP packages, all material flow recording is based on unique part numbers. In information systems for engineer-to-order or make-to-order production, the situation is definitely different. In these information systems, all materials other than standard items are stored and recorded by work order. To some readers, this may wrongly appear to be a minor point. Therefore, we shall outline some consequences of this difference.

First of all, all inventory applications in MRP packages are based on part numbers. For example, cycle counting, physical storage allocation, inventory evaluation for the balance sheet, supplier performance evaluation, turn-over rate estimation, computation of storage costs, to mention only a few. All these applications have to be expanded considerably when inventory may become customer specific.

Secondly, budgeting applications, costing applications and applications supporting marketing and profitability analysis are in MRP packages based on code numbers. Again, the whole application has to be redesigned when customer orders no longer refer to code numbers.

Finally, life-cycle support of products such as service contracts and guarantees are based on code numbers. When code numbers are no longer used, other structures such as the generic bills of material (see Chapter 26) are required, but these structures have considerable impact on the nature of the application software.

27.3.5 Decision support

The decision structure presented in Chapter 21 is seen as the core of the control system for customer driven manufacturing. Within this structure, five decisions at the GFC level and four decisions at the PUC level have been identified:

- Evaluating and selecting requests for quotation (GFC);
- Issuing quotations, pricing and due date assignment (GFC);
- Work order release for the Product Engineering PU (PUC);
- Work sequencing for the Product Engineering PU (PUC);
- Reserving resource capacity within the Detail Design PU (GFC);
- Internal order acceptance (GFC);
- Work order release for the Detail Design PU (PUC)
- Work sequencing for the Detail Design PU (PUC)
- Sub-order assignment and outsourcing (GFC).

A number of decision support systems can be identified in the outermost layer of the information architecture which support the above-mentioned decision processes. Although a broad range of planning techniques exist, of which a few also appear to be suitable for use for customer order driven engineering as described in the selected case situation, much of the data is too general and unreliable in this particular situation. This means that it would be more effective to invest in creating *slack* and *alertness* rather than spending time and effort on optimal planning.

Alertness is needed in situations in which there is a large amount of uncertainty; if problems cannot be adequately anticipated, then the alternative is to react quickly when problems occur. A certain amount of slack needs to be arranged in advance in order to allow for effective reactions. A good decision support system can improve the alertness and speed of decision-making and communication. This means that the correct information must be provided to the decision maker as quickly as possible. The decision support systems need to be designed with this in mind. The various decision support systems are included in the outermost layer of the information architecture diagrammed in figure 27.8.

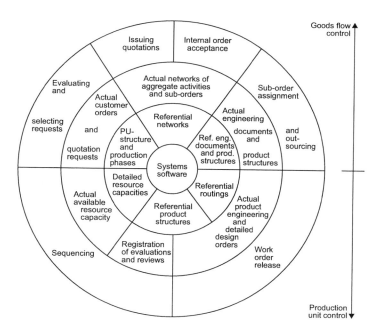

Fig. 27.8 An information architecture for customer order driven engineering

27.4 CONCLUSION

What can be concluded from the foregoing discussion?

First of all, it should be clear by now that planning based on the material structure of the products differs from planning based on gradually emerging activities in networks. It is often impractical or wrong to consider the product structure (bill of material) as the basis for activity planning. This means that software which is strictly based on MRP (or OPT) is not generally applicable in engineer-to-order.

Secondly, engineer-to-order production can be interpreted as a sequence of main stages, where each stage provides for specification of the detailed activities of the subsequent stage. This view differs from the generalized multi-project planning situation where all data are assumed to be available.

Thirdly, the different nature of resources in customer-driven manufacturing as discussed in Chapter 17 should be reflected in information systems. Modelling human skills and capabilities of equipment should be enabled.

Finally, the decision structure as presented in Chapter 21 can be supported by information systems if these are carefully designed with the application of engineer to order in mind.

REFERENCES

1. Betrand, J.W.M. and J.C. Wortmann, *Production control and information systems for component manufacturing shops*, Elsevier, Amsterdam, 1981
2. Bertrand, J.W.M., J.C. Wortmann and J. Wijngaard, *Production Control. A Structural and Design Oriented Approach*, Elsevier, Amsterdam, 1990.
3. Martin, J., *Data structure diagramming techniques*, Prentice-Hall, 1986
4. Wight, O.W.W. and D. Landvater, *The Standard System*, Manufacturing Software Inc., Williston, Vt, 1983

28

Document Management in Customer Driven Manufacturing

28.1 INTRODUCTION

The principles of data management presented in the previous chapter are used as a basis for discussing the quality of document management in this chapter. Data management deals with the *contents* of documents whereas document management is concerned with controlling documents at the lowest level of detail, regardless of the contents of these documents. This chapter covers the procedural aspects of document management in customer driven manufacturing and provides insights into how information technology could be used to support these procedural aspects.

Document management should be interpreted in the broadest possible sense within the context here. Document management, thus, may incorporate:

- change management to control the process of influencing a product, a process or the organisation that, in turn, affects the documentation;
- work flow management that may lead to new information requirements;
- managing the reuse of information and limiting diversity.

The methods for dealing with different types of data that were presented in Chapter 27 are used in this chapter. In addition, the concepts introduced in Chapter 26 are applied here to describe a more general approach to using generic bills of material for different kinds of documentation.

The purpose of this chapter is to explain why document management is a major issue in managing uncertainty and complexity in customer driven manufacturing. It is also explained how a structured way of dealing with documents leads to decreased time to market and more structured, simplified procedures.

The following topics are covered in the remainder of this chapter:

- the trends in document management,
- the activities needed to perform document management effectively,
- the specific benefits provided by document management activities in customer driven environments, and
- the role of a document management system in supporting these activities.

28.2 TRENDS IN DOCUMENT MANAGEMENT

The importance of data management was discussed in the previous chapter. However, even in situations where accurate data management principles are followed strictly, a number of problems related to document management may arise. The following trends can be seen as the source of these problems and the reason these problems are becoming more prevalent in today's customer driven manufacturing organisations:

- Product specifications are being determined more frequently by the market situation. This leads to a broader range of product variants and more customer-specific information. As a result, a greater number of documents and document types must be managed;
- The complexity of products is increasing, mainly due to information technology and miniaturisation. This increased complexity of products leads to an increased complexity of the related documentation;
- More concurrent engineering is necessary to reduce time to market. This creates the requirement for a better and faster exchange of information and the need for an improved management of document versions and document status;
- The increasing use of product families (see also Chapter 26) requires product family documentation and the ability to easily create documentation for specific situations;

- The increasing demand for reusable ideas, solutions, rules and constraints implies a need for documenting these issues and enabling intelligent but easy-to-use search functions;
- Companies are increasingly obliged to follow quality assurance procedures where documentation is concerned (ISO 9000 series of standards). Improved procedures for documentation management are required.

The combination of these trends is leading to an increasing number of problems with regard to the complexity and size of the document management effort required within customer driven manufacturing companies.

28.3 DOCUMENT MANAGEMENT ACTIVITIES

Before going into detail on the aspects to be controlled in handling the complexity and size of document management activities, it is necessary to define what a document is and what document management means.

A *document* in this context is *a collection of data being communicated and managed as a single unit.* Examples of documents are CAD drawings, test instructions, wiring diagrams, mountings, photographs, bills of material and quality manuals. This means that a document is not just data on a piece of paper or in a file, but rather a collection of related data on any kind of medium.

Document management is defined as *the collection of all of the activities related to creating, retrieving, delivering, changing, deleting, and maintaining documents in such a way that the right document is available to the right person at the right place and time.*

For achieving optimal document management it is necessary to define which *activities* are included within the scope of document management and which *meta-data* elements are essential to support decisions regarding these activities. The decisions to be made typically concern issues such as who created the document, which persons are using it, who has a copy, who needs the document at which point in time, who may use it for what purpose, what relationships exist between the document and other documents, products, processes etc. The meta-data elements related to document management include the document version, status, owner, creation/modification dates, current users, type, access rights and relationships with other documents.

The above definition of document management implies that (parts of) the following activities are related to document management:

- Document Control;
- Change Management;
- Supporting Reuse;
- Generic Document Management;
- Work Flow Management.

Figure 28.1 illustrates how the meta-data elements are related to a document and how the aforementioned activities are related to a document during different phases. These activities are described in the following subsections and explained in more detail in Section 28.4 in connection with the benefits of document management in a customer driven manufacturing environment.

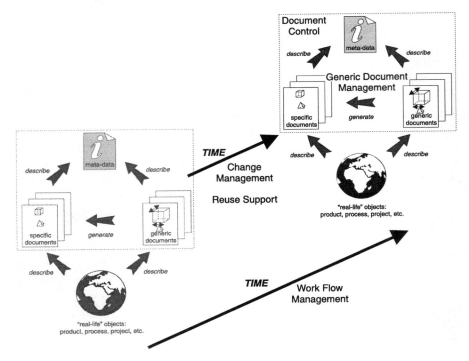

Figure 28.1 Documents, meta-data and document management activities

1. Document Control

Document control is an activity that is concerned solely with document management and involves the basic functions of accessing, changing, storing, and maintaining documents. The *accessing* function supports finding the document, retrieving it and presenting it to the user based upon the appropriate "view." *Changing and storing* refers to the physical changing and storing of a document including change and storage rights, classification of the document, delivery of the document to subscribers after modification and similar functions. Note that the procedural aspects of making changes (e.g., deciding whether to change something, the implications of making a change, updating the version number and status) fall under Change Management (covered in the next section). *Maintaining* documents refers to ensuring that documents are available, regardless of the version of a program or tool being used or the particular user interface employed. This function also includes ensuring document recovery after hardware or software crashes.

Another important part of document control is maintaining meta-data elements. The meta-data elements are used and sometimes changed in performing actions on documents. The document relation structure is a key component of the meta data and is stored as a part of document control. This structure defines the relationships between documents and other objects. These objects may be related documents, actual products, processes, projects or other entities involved in the document management procedures. The document relation structure is used by most of the other document management activities since it provides the necessary links between documents and the network structure needed to locate and reuse documents.

2. Change Management

Change management is concerned with supporting the user throughout the total change process: guiding the user in the steps to be taken during the change process, creating possibilities for impact analysis, registering change requests, orders and decisions and managing the changes in version and status.

A change process may involve a minor change or a major change, depending upon the importance of the change, the type of data involved (customer-dependent or independent) and the level of the data involved (operational, tactical, strategic). Despite whether a change is major or minor, however, a change process always consists of the same steps, in principle. These steps could be carried out as part of a single individual's thought process in the case of a minor change. The change process steps are as follows:

- define the change proposition (contents, owner, date, reasons for change);
- perform an impact analysis by checking which objects are influenced by the change (e.g., documents, products, processes). This can be accomplished based upon the previously mentioned document relation structure and other factors;
- solicit the advice of all of the persons responsible for objects that are affected by the proposed change;
- decide whether to make the change; document and communicate this decision;
- perform the actual change with respect to all of the affected objects;
- review and approve all changed documents, including changes to the relevant version and status information (also called "design release management").

All these steps need to be followed implicitly or explicitly to ensure that changes in documentation are made reliably. It is important to note that the aforementioned steps will generally be carried out as major activities involving a multidisciplinary group when the change proposal involves customer-independent documentation. The situation is normally different when only customer-dependent data elements are involved; only minor change processes are typically proposed in this case and the activities will then be carried out by a limited number of individuals.

3. Supporting Reuse

The third aspect of document management concerns reusing the information stored in documents. The reuse of information typically refers to previously formulated ideas, solutions or infeasibilities involving product components, modules, orders, projects and similar objects. Different forms of reuse may be found, depending upon the type of company (e.g., assemble-to-order, engineer-to-order), the phase in the product life cycle (e.g., development, production) and the type of activity.

Particularly related to document management, the main problems encountered in reusing information are:

- finding the appropriate information. The typical search functions provided in standard document management applications only support simple search and browse commands such as "look for the term X" or "display page Y of document Z." This is not sufficient for quick, easy and intuitive information searching;
- communicating and transferring the information. When the information to be reused is known but must be accessed by different users on different hardware and software platforms, data communication and performance problems may be encountered;
- being able to reuse and extract only the relevant parts of the information contained in a document. A complete document is rarely reused in its entirety. Generally only certain document parts, sections, or objects are reused. It is often difficult to extract only the relevant information from an existing document and include it in a new document.

In practice, designers will be receptive to the idea of reusing information only if these three problems are resolved to a sufficient extent. Otherwise, designers will prefer duplication of the information in a new application. Furthermore, if no allowances are made for the potential reuse of information in the system design and the storage of data, then the actual possibilities for reuse of information may be limited.

To resolve the aforementioned problems in reusing information, the information needs to be stored initially in such a way that all of the meta-data elements needed to support reuse are included. These are related to:

- classification and standardisation. Related objects stored in separate documents need to be included in logically-defined groups that are intuitive for the designers;
- association and browsing. When storing information it should be possible to relate documents to associated information such as the document source. In subsequent searches, it should be possible to retrieve the associated information in an easy way. The most appropriate techniques for this often utilise hypertext and hypermedia features;
- annotation. It is not sufficient to provide just the capability of easy access to ensure the information is reusable. The degree of reusability also depends on knowing what is used or not used, which instances

are relevant and which reasons are given. To achieve this, a document management system should provide a free-format annotation capability for establishing links to all documents, relationships and meta-data elements.

4. Generic Document Management

The product structure approach used mainly in assemble-to-order companies was described in Chapter 26. The generic bills-of-material approach was proposed to support the concept of product families in this type of situation. A generic approach can be used for other purposes other than just the bills of material. In is clear that other documentation like routings, drawings, price calculations and assembly instructions can be handled at the family level. This means that for all these types of documents a description of a family document can be produced for the purpose of generating a specific document for a specific situation. This approach should also be seen as a part of document management because generic documents require:

- different document storage structures;
- different creation and delivery approaches for specific documents;
- different change management approaches.

Generic document management is therefore concerned with supporting the user in recording and maintaining a product family and process structures together with related documents, keeping in mind different user views (see also Chapter 26).

In dealing with generic family documentation, the descriptions of several specific documents are combined into a single generic document. The information stored in the generic document, which is subsequently selected as appropriate for a specific order, can also be seen as an effective way of reusing information. Since generic document management is such a powerful concept, however, and goes much further than other examples of reusing information, it is treated here as a separate document management activity.

5. Work Flow Management

Work flow management deals with guiding and supporting users in defining, planning, registering and monitoring processes from the point of view of managing all of the interrelated information flows. This implies that work flow management is not really an integral part of document management, but should be regarded as a separate activity. Work flow management is, however, strongly related to document management. In the event that certain documentation is required as input for an activity, it is obvious that any documentation which arrives too late can be seen as a critical element. However, it is also important to note that it may be undesirable in many cases for documentation to be available too early. In such cases the information may not be received (because the receiver does not know what to do with it) or it may be disregarded (if the information is incomplete or the receiver is not able to process it). This means that the rule of having the right information available at the right time and place should be strictly observed. This can only be done by having a well-organised work flow management function for the management of documents.

The above situation (in which information is available just-in-time for use as input for an activity) can be compared to a situation in which goods or components need to be available at the right time and place for certain production activities. This suggests that techniques used for production control involving the flow of material may also be applicable to production control of information flows (see also Chapters 15 and 23). Examples of such techniques include batching, JIT and network planning.

28.4 BENEFITS OF DOCUMENT MANAGEMENT IN CUSTOMER DRIVEN MANUFACTURING

Based upon the general descriptions of the document management activities in the previous section, several specific aspects of these activities in customer driven manufacturing environments are described in this section.

1. Document Control

As discussed in the previous section, document control deals with accessing, delivering, changing, storing, and maintaining documents. These functions are similar in virtually every type of company environment and, therefore, will not be discussed extensively here.

Nevertheless, it is important to point out that for customer driven manufacturing (parts of) the documentation need to be created, generated or designed after the customer order has been received. This means that providing the necessary documentation becomes one of the bottlenecks in efforts to reduce the lead time. In assemble-to-order situations this problem can be solved by using a product family approach (see Generic Document Management and Chapter 16), but in engineer-to-order situations the timely availability of documentation is often a serious problem.

In these cases, lead times can be reduced by accelerating the development and improving the design of product documentation. Lead times can also be reduced by enabling faster and easier communication between all of the disciplines involved. Faster and easier communications are typically realized through:

- the use of document management systems that provide the most recent version of the right document at any time and any place. New versions of documents can be distributed automatically or notifications of updates can be distributed to all subscribers or users of an updated document;
- simulating the work flow activities using a model to determine which activities require which documentation. This means that a document can be completed and provided to a specific activity based upon the just-in-time principle. If the document is still being prepared, then a notification can be sent to the waiting activity to eliminate the need for time-consuming searches. In this situation, the document can then be delivered to the waiting activity when it is ready;
- the use of a document management system together with a work flow management system to provide a clear view of the completed activities, activity outputs, current activities and the documents required by these activities.

2. Change Management

As discussed in Section 28.3, Change Management involves supporting the users during change processes. A number of factors need to be considered with respect to managing changes to documentation in customer driven manufacturing situations:

- The number of document versions is relatively high because the customer-independent documentation is supplemented with customer-dependent data for virtually every (type of) document and every order.
- Lead times must be kept short and concurrent engineering methods are used, thereby increasing the complexity of version and status management.

Benefits of document management in customer driven manufacturing 367

- Totally different approaches should be followed for managing changes to customer-dependent documentation and changes to customer-independent documentation. The different approaches to Change Management are described in Section 28.3.

Improving the management of changes in concurrent engineering environments can, of course, be accomplished by improving communication as mentioned under the Subsection 1, above. However, the main problems in concurrent engineering environments involve the integrity of data stored in documents. Because several individuals may be working on the same subject simultaneously, the same documents may be in use by several people. The usual way of dealing with this requirement for concurrent use is to:

- give read/write access privileges to the first person to access a given document and temporarily lock-out any other users that request read/write access to the most recent version of this document by changing the document status from "released" to "busy";
- ensure that all other users using the same document can only get a read-only copy of the version stored in the data base. Read-only users are not allowed to make any changes to the original document;
- change the document status back to "released" when the first user accessing the document has finished modifying the document. All other users working with the read-only copy of the old version are then notified that a new version is available. Another user can now access the updated document with read/write privileges to make additional changes as necessary.

The process described above is used in most concurrent engineering environments but results in forcing developers to wait frequently for other developers to complete their work. A better approach could be implemented as follows:

- allow all users to access a given document, simultaneously, with read/write privileges. The document is assigned the status of "busy" while it is being accessed and used;
- when one of the users is finished with making changes to the document, the version number of this document is compared with the latest version of the document stored in the data base. If the version numbers of the two documents are identical, then a new

version number is assigned to the updated document and all users of the document are notified of the change and the existence of a new version;
- if the version number associated with the updated document is lower than the last version of this document stored in the data base, then clearly some user has already made changes to the document in the meantime. A check then needs to be made to determine if the latest updates can be made in the newer version of the document stored in the data base. If this is possible, then a newer version of the document is created which is consistent with all of the changes.

The advantage of this approach is the elimination of the need to wait for a concurrent user to complete changes to a document before another user can process additional changes. This approach is particularly useful when different parts of a single document are changed by specific designers. Problems can only occur when a changed document needs to be checked and changed again after a blocked attempt to update a newer version of this document in the data base when there are concurrent users. Nevertheless, the waiting times for physically transporting a document from one user to the next user are likely to be much greater than the time needed to check and reprocess changes when a document is updated concurrently.

3. Reuse Support

Within the context of document management, the reuse of information mainly involves supporting users in classifying and finding previously used designs or documents to limit unnecessary diversity and design efforts. If provisions are made for searching for this design information at the time that it is stored in the data base, then the design lead times can be reduced. Reuse is, thus, often useful in customer driven manufacturing organisations.

The actual degree to which documentation is reusable and the type of reuse depends largely upon the type of organisation. In assemble-to-order situations, two types of reuse can be distinguished. The first type is the reuse of knowledge (represented in documentation) in handling a specific customer order. With this type of reuse the items to be reused are totally predictable and can, therefore, be included in an intelligent model (see Generic Document Management). The second type of reuse is the reuse of ideas, solutions, components, assemblies etc. in product families.

Engineer-to-order situations are similar with respect to the types of reuse. The first type of reuse here deals with variants of parts that are always incorporated in the manufactured products. The second type involves the parts to be engineered. Compared to the assemble-to-order situation, however, the second type of reuse is much more common. For the second type, a classification can be made based upon the characteristics of the items to be reused [Muntslag, 1993]:

- engineering (and therefore reuse) based upon a specific technology;
- engineering based upon pre-defined product families;
- engineering based upon pre-defined product sub-functions and solution principles;
- engineering based upon pre-defined product modules;
- engineering based upon pre-defined finished goods.

Depending upon the level at which engineering is required, different design and reuse items may be applicable. The degree of reuse can be increased by placing extra emphasis on a proper classification scheme and standardising the documentation of these items (see Section 28.3).

4. Generic Document Management

Generic document management involves managing the information stored in generic documents and the generation and distribution of specific documents related to a generic document. Generic documents typically consist of a full description of:

- all possible characteristics and options (parameters and parameter values) which could uniquely describe a specific variant of the family document;
- all of the constraints that restrict the number of feasible combinations of these parameters and parameter values;
- all of the components that may be included as part of a specific document, together with the combinations of parameters and parameter values that are valid for each component.

In customer driven manufacturing situations where generic documents can be described in this way, such documents can be used to reduce the volume of specific documents that would otherwise need to be stored, kept and maintained. Depending upon factors such as the modularity and number of modules or components, a single generic document may replace anywhere from 100 to 100,000 specific documents. Besides the advantage of less maintenance, there are other advantages to using generic documents:

- the strict separation of state-dependent and state-independent data. It is also possible to automatically maintain the relationships between these two types of data stored in the document relation structure;
- improved possibilities for retrieving specific documents by specifying parameters and parameter values more intuitively in the "language" of the users of these documents. In this way the generation of a specific document can be seen as an optimal way of reusing stored knowledge.

5. Work Flow Management

In section 28.3 it was made clear that work flow management is not part of document management. However, the relationship between work flow management and document management is an important one. Work flow management itself will for that reason not be detailed here. Only the relationship between activities and documents will be looked at.

In assemble-to-order situations the time and capacity needed for the assembly of products for a specific order are normally critical. Modelling the work flow for generating the documentation is most useful in this situation. Since all steps for a specific order can be predicted in an assemble-to-order situation, a full model can be developed for the activities to be performed. Such a model can even be developed in the form of generic documentation that can then be used to generate the specific model for a customer order. The kind of documentation necessary to perform the individual activities can also be described in advance. For a specific order it is then easy to create a quick overview of which activities need to be performed at which times; subsequently, the documentation necessary to handle each activity can be generated exactly at the point in time when it is needed.

An engineer-to-order situation is more difficult. The time and capacity are also critical factors during the design phase. Certain parts of the design and production activities can be handled in the manner described above. In this case there is only one way to handle the uncertainty: by starting with a general description of the work flow and related documentation. The amount of uncertainty subsequently diminishes in each successive design step as more of the description details are worked out (see also Chapters 14 and 29). The result is that most of the documentation input requirements are specified and ready by the time that the documentation is needed to carry out a specific activity.

28.5 THE ROLE OF A DOCUMENT MANAGEMENT SYSTEM

The description of the various functions associated with document management provides a basis for evaluating how information technology can be utilised to support these functions. The role of a document management system is discussed in this section. Special attention is paid to how such a system can be used to support users, applications and documents.

Clearly the data elements concerning products, processes and projects are critical to the business. This type of data will typically be retrieved, used and changed by many individuals who may be working concurrently. Furthermore, there may be different versions of a document and the status of a document may determine the access rights and way in which it is stored. Managing these complex aspects implies that a document management system should control *all* product, process and project data within a company.

The best way to promote the use of standard procedures and ensure complete data integrity is to have a document management system as the *only* entity with access to product, process and project data. In this way, the document management system can be used to ensure that users are not permitted to access data directly, but only through the controlled functions of the document management system. This implies that the document management system must keep track of which users have access rights to which version of certain documents and what these users are allowed to do with documents based upon characteristics such as the status and type of document. This requires that information be collected and maintained about the data stored in the documents. This information was referred to, above, as meta-data. The meta-data is generally stored in one place and forms the core, the main source of information, for the document management system. The normal data contained in documents (i.e., data that is not meta-data) can be stored at various locations in any number of data bases throughout the organisation, regardless of whether these data bases are automated.

Figure 28.2 shows how an advanced document management system is related to document data and meta-data. The figure also shows that document management is only concerned with the procedural aspects of dealing with documents and not with the technical aspects of a document being stored, found or retrieved. The technical aspects of data management are addressed in the implementations of (standard) data base management systems (DBMS).

Figure 28.2 also shows how users and applications are related to a document management system and the documents, themselves. The document management system serves as an interface between the users and applications on one side and documents on the other. This is a valuable aspect from the user's point of view since it provides a natural point for controlling user access. As a result, unauthorised users are not able to jeopardise the integrity of the data. Safeguards can be included to ensure that the right document version is used, data is stored appropriately, etc. When required, a document management system can also limit access to documents for specific users (e.g., read-only access with no capability for changing, creating or deleting data).

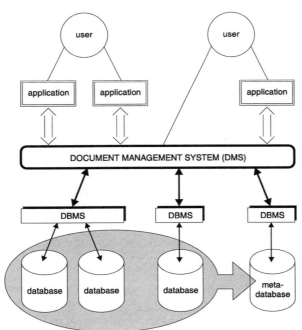

Figure 28.2 The role of a document management system

A situation in which the document management system provides full access control between the data and the rest of the world is rare, however, in modern systems. Full access control implies that none of the application programs would be able to circumvent the controls imposed by the documentation management layer in accessing, changing and storing data. Production control systems, CAD systems, word processing applications, etc. would all be controlled by the document management system when finding, retrieving and storing data files. This implies that special interfaces would need to be developed (e.g., using a macro language) to allow all of the applications to communicate with the document management system at this level.

Document management systems that provide partial coverage of the above-mentioned functions include WorkManager (Hewlett Packard), ProductManager (IBM), InfoManager (EDS) and DMS (Sherpa). All these systems run on several platforms, support various DBMS's and include features for modelling activities. Because of the work flow management aspects related to document management, these systems are also known as Product Data Management, Engineering Data Management or Work Flow Management software packages. Most modern document management systems available today provide only a limited number of functions and features, however. They are generally restricted to managing documents on a single platform, may only support a single application and typically provide only simple store, search and retrieval functions as an isolated application.

29

IT-support of customer-driven engineering management

29.1 INTRODUCTION

Management of tendering and engineering has been discussed in Chapter 22. This chapter showed how engineering activities can be considered as normal workflows with normal management and control procedures. Chapter 22 has elaborated on the case of the company producing packaging equipment that has been described in Chapter 9.

One of the conclusions of Chapter 22 is, that there should be a clear distinction between the quotation phase and the order phase. These phases are in this case handled by the production units (or better: engineering units) "Product Engineering" and "Detailed Design" respectively (cf. fig. 22.1). A similar situation has been described for the shipyard of Chapter 12, where these phases are called "bid preparation" and "design". Progress control and other management activities within these units is called *Production Unit Control* (PUC), in order to stay consistent with the remainder of this book.

In Chapter 27 we have presented a general architecture and data structure for customer-driven manufacturing. The discussion in Chapter 27 was not focussed specifically on design and engineering. In the present chapter, our focus will be on design and engineering.
According to Chapter 27, information systems should make a distinction between state-independent and state-dependent data. For both types of data, product descriptions, workflow descriptions and resource descriptions should be allowed. This will be elaborated in the present Chapter for the engineering activities and for engineering management. The products of these engineering activities are documents, in various forms, as has been described in Chapter 14.

376   *IT-support of customer-driven engineering management*

Requests for quotation and customer orders should be registered and monitored at PUC level. This is accomplished via registration of Product Engineering orders and Detailed Design orders, in the state-dependent layer of figure 27.8. The (planned and actual) resource capacity requirements of these orders can be used to determine the actual resource capacity utilization (time phased). The registration of the results of technical evaluation and the various engineering reviews used for quality assurance is also found within the state-dependent layer. The aforementioned data elements are included in the state-dependent layer of the information architecture as illustrated in Figure 27.8.

29.2 THE STATE-INDEPENDENT DATA MODEL

The complete state-independent data model for controlling customer order driven engineering in the selected case of Chapter 9 can be split into product-oriented data structures and workflow-oriented data structures and resource-oriented data structures as follows:

- reference product data and standard product data
- reference routings and networks of detailed activities (workflow data at PUC level)
- reference networks of aggregate activities (workflow data at the GFC level)
- standard resource data.

The last two kinds of data structures have been discussed sufficiently in Chapter 27. The first two points require some elaboration for the engineering activities. These data structures are presented and explained separately here, for the purpose of clarity. The complete state-independent data structure is then presented as a whole.

29.2.1 State-independent product data

The state-independent product data (see figure 29.1) represents the results of the customer-order independent (or "innovative") engineering. This figure is an extension of figure 27.6. A number of *reference final products* are defined in a general sense or, in some instances, in detail for each *product family* based upon the expected market

The state-independent data model

developments and requirements. These product descriptions are then used as reference models to support the product engineer and detail designer in their efforts to develop a customized product. The degree to which the reference final products are developed and documented independently from any specific customer order may vary from one or more entities of type *reference sub functions* to complete *reference product modules* which are fully defined at the *item* level.

N.B. Product engineers will generally redevelop components and modules if they discover that an excessive amount of time or effort is required to find the specifications for a reference product. The *generic bill-of-materials* principle (Chapter 26) can be seen as an important development with respect to the representation of product structures and could provide valuable support in this context. The way in which this type of principle could be used effectively within customer order driven engineering is not included within the scope of this chapter, however.

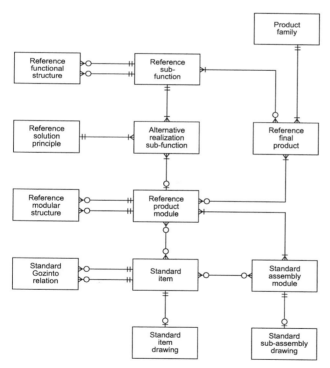

Fig. 29.1 State-independent product data

29.2.2 State-independent PUC-workflow data (routings and detailed activities)

Detailed reference data concerning the customer-driven engineering process is maintained in addition to the required aggregate workflow data. More specifically, the workflow that is required for control at the PUC level should not be re-documented for each new customer order (see Figure 29.2).

Various *production units* (PUs) can be identified within customer order driven engineering from a control point of view. The data structure which represents these production units is state independent. One or more *reference production phases* are carried out within each PU. Each entity of type *reference production phase* is comprised of one or more *reference product engineering/detailed design operations*. For each entity of type *reference operation*, the required product engineering/detailed design resource capacity is specified. The output of a reference production phase is a *reference engineering document*. An entity of type *reference engineering document* represents the product description to the extent that it has been developed at a given stage of the customer order driven engineering process.

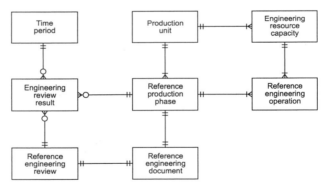

Fig. 29.2 State-independent PUC-workflow data

In addition to data concerning the workflow, a number of state-independent reference data elements which deal with quality are also maintained. A *reference engineering review* is defined for each reference engineering document. A number of engineering reviews are performed at different points in time within each production phase for the purpose of quality assurance as an integral part of the processing of various customer orders. The results of these reviews (in terms of

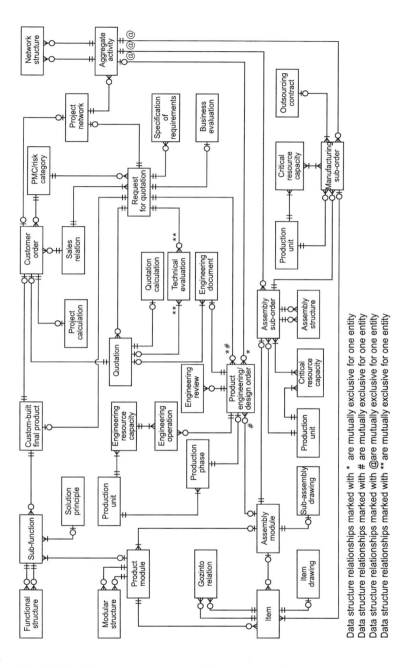

Fig. 29.3 The state-dependent data structure diagram

the total number of faults) are registered as state-independent values per production phase per period. A periodic analysis of the engineering process can be prepared (to control the quality of the process) based upon this data. As time progresses, the quality of the process can be evaluated based upon these quality data.

29.3 THE STATE-DEPENDENT DATA MODEL

The complete state-dependent data model for the control of customer order driven engineering in the selected case situation can be split into product-oriented data structures and workflow-oriented data structures (analogous to the state-independent data structure) as follows:

- state-dependent product data
- state-dependent aggregate networks and order structures (GFC level)
- state-dependent process data (PUC level).

The state-dependent data structure diagram concerns the registration of all of the quotation-dependent and customer-dependent product data and process data. The *requests for quotation* in the quotation phase and the *customer orders* in the order phase are of primary importance here. In contrast with the way in which the state-independent data structure diagram was described, the state-dependent data structure diagram will not be split into sub-structures here. Many of the entities and relationships between these entities are analogous to the entities and relationships found in the state-independent data structure, with the major difference being that the data here is customer-dependent. The complete state-dependent data structure diagram is presented in Figure 29.3. The most important entities and their relationships are described chronologically according to the creation and the subsequent processing of a customer order.

The customer order driven engineering process starts with the arrival of a *request for quotation* from a *sales relation*. A significant part of the activities are closely related to this request. A large amount of data is similarly linked to a request. The sales organization documents the customer's *specification of requirements* to support and clarify the request for quotation. A *business evaluation* and a *technical evaluation* are prepared for the purpose of Evaluation and Selection. More than one technical evaluation may be performed if multiple quotations are issued or if the specifications are changed during the quotation phase. If a request for quotation is accepted, a *product engineering order* is issued for preparing the quotation details. The product engineering order (which consists of a single *production phase*) is carried out within the *Product Engineering PU* in the form of a number of *engineering operations* and the use of a certain amount of *engineering resource capacity*. The product engineering order is comprised of instructions for the preparation of an *engineering document* which includes the preliminary draft of the future product. The engineering document, thus, represents an intermediate form of the final product. The *functional structure* and the *modular structure* which comprise the contents of this document are documented separately for each request for quotation (see Figure 29.3). As such, a distinction is made between the product which is required as the result of the product engineering order (the aforementioned engineering document) and the detailed preparation and documentation of the product description (the aforementioned functional structure, etc.). An engineering document can be seen as a description of a final product at a certain stage of development. The engineering document, as such, does not change, however; only the level of specification of the final product under development changes as the subsequent detailed design orders are completed during the order phase. The various steps in the engineering process can be reviewed based upon the sequence of engineering documents which are prepared, registering the status of the product description at each stage. The data structure related to this has been excluded from Figure 29.3 in order to improve the clarity of this diagram.

It is possible that various successive quotations are issued during the quotation negotiations. This may also lead to successive versions of engineering documents and multiple engineering orders. A *quotation calculation* is prepared as the basis for the quotation price. A *project network* consisting of a number of *aggregate activities* with an associated *network structure* is defined in order to determine the throughput time for the quotation.

The quotation may lead to a *customer order* in some instances. Each request for quotation and each customer order is classified in a *PMC risk category*. The original project network is subsequently linked to the customer order. The quotation calculation is translated into a *project calculation*. In addition, the final product under development in the quotation phase is linked to the customer order and then developed in more detail. Performing the aggregate activity of Detail Design leads to various detailed design orders for the further development of the product description within the Detail Design PU. The total product is divided into *assembly modules* before the detailed design activities are initiated.

The aggregate activity of Assembly can be detailed further in the form of an *assembly structure* consisting of *assembly sub-orders* as soon as the assembly modules are defined. The various detailed design orders have an important relationship with the different assembly modules. Some of the detailed design orders are related to the general detailing of an assembly module while other detailed design orders are related to the detailed development of the different components within an assembly module. It may be decided to conduct one or more *engineering reviews* for each product engineering/detailed design order to evaluate the quality of the product description at certain points in time. The aggregate activity of Component Manufacturing is detailed further in the form of *manufacturing sub-orders* during the detailed design of the product. A manufacturing sub-order consists of instructions for producing a group of *items* related to a specific assembly module which is produced in the same *production unit* during approximately the same period of time. An estimate of the required amount of critical resource capacity is made for each manufacturing sub-order to be used as a basis for the evaluation of the capacity in connection with Sub-order Assignment and outsourcing decisions. One or more of the manufacturing sub-orders which are to be sub-contracted to the same supplier may be combined in a single *outsourcing contract*.

29.4 THE RELATIONSHIPS BETWEEN THE DATA MODELS

The state-independent and state-dependent data models were discussed separately in the preceding sections. Nevertheless, these substructures should be seen as parts of a single, total data model in which important relationships between these two structures can be identified. The connections and relationships between the state-independent and state-dependent data models are described further in this section. For the sake of clarity, the complete data model will not be presented as one diagram since the large number of relationships between the state-independent and state-dependent data structures is rather overwhelming. The relationships between these data structures can be explained more adequately by presenting them separately in terms of the relationships between:

- reference product data structures and customer-specific product data structures;
- reference workflow data structures and customer-specific workflow data structures.

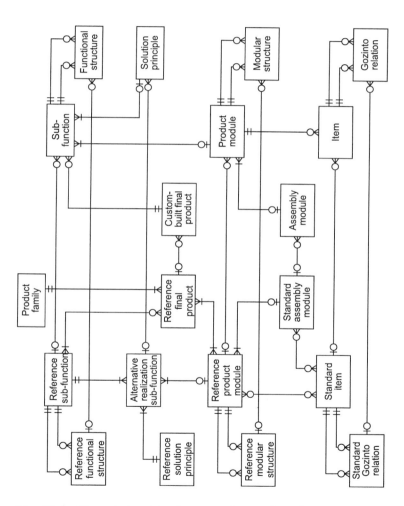

Fig. 29.4 The relationships between the product data structures

384 *IT-support of customer-driven engineering management*

The reference product data such as reference sub-functions and reference product modules should be used by the product engineer and the detail designer as initial starting points for developing customized products. From the point-of-view of control, it is important that analyses can be made regarding how and how frequently that reference product data are used for the development of customized products. This means that the registration of the use of available reference product data for developing customized products should be included in the data model (see figure 29.4). The registration of which reference product data has been used and when it was used, is performed per engineering phase, starting with the *custom-built final product*, including the *functional structure* and ending with the *modular structure*.

A similar line of reasoning can be followed for the use of reference workflow structures. These relationships are illustrated in Figure 29.5.

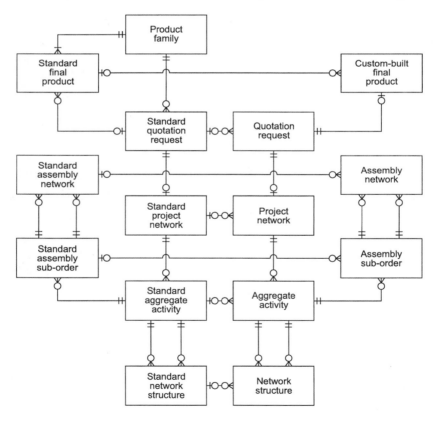

Fig. 29.5 The relationships between the GFC-workflow data structures

30

Production Unit Control

We shall now consider the information system used to plan and control the flow of products in any particular production unit. Each supervisor has waiting on his/her desk a list of customer requirements, which have to be fulfilled for the forthcoming week. The term "customer" may refer to an external customer or to another production unit in the same company. The main task facing the supervisor at this moment in time is to plan production over the following working week to ensure that the customer orders will be fulfilled. Factors which will influence the content of this plan include the likely availability of resources (operators, machines) and the capacity of the manufacturing system. This plan will then act as a reference point for production, and almost certainly will have to be changed due to any number of unpredictable events arising (e.g. raw material shortage, operator problems, quality problems, or machine breakdown). Therefore, the three main elements for shop floor control are:

1. To develop a plan based on timely knowledge and data which will ensure all the production requirements are fulfilled. This is termed *scheduling*.
2. To implement that plan taking into account the current status of the production system. This is termed as *dispatching*.
3. To *monitor* the status of vital components in the system during the dispatching activity, either with the naked eye or by using technology based methods.

The activities of scheduling, dispatching and monitoring are in fact carried out, perhaps informally, by every competent shop floor manager or supervisor. Our approach is to outline formally each of these separate tasks and show how they interact to control a manufacturing system.

Production Unit Control

The basis of our approach is to map the various shop floor activities onto an *architecture* which recognizes each individual component. The advantage of having an architecture is that it both formalizes and simplifies the understanding of what occurs during production, by establishing clear and separate functions which combine into a shop floor control system.

PUC realizes the lowest level of the PMS hierarchy. We will follow here the COSIMA project which recognized the need for a generic and flexible architecture which identified and separated the different functions of PUC.

The architecture is illustrated in Fig. 30.1, and the five basic building blocks of the PUC system are the *scheduler, dispatcher, monitor, mover* and *producer*. The *scheduler* develops a plan over a specified time period, based on the manufacturing data and the schedule guidelines from the Goods Flow Control level, GFC. This plan is then implemented by the remaining four modules. The *dispatcher* takes the schedule and issues relevant commands to the movers and producers, which carry out the required operation steps necessary to produce the different products. Given our inability to accurately predict the future, the need to modify the plan due to unforeseen circumstances (e.g. machine breakdown) may arise, and in this case the *monitor* notifies the dispatcher of any disturbances. The schedule may then be revamped to take account of any changes in the manufacturing environment.

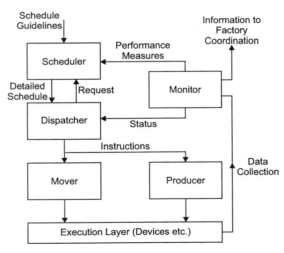

Fig. 30.1 Production Activity Control

Based on the instructions from the dispatcher, the *producer* controls the execution of the various operations at each workstation. The *mover* organizes the handling of materials between workstations within a cell by following the dispatcher's commands.

The scheduler

The task of the scheduler is to accept the production requirements from a higher planning system (i.e. a GFC system), and develop a detailed plan which determines the precise use of the different manufacturing facilities within a specified time frame. Good scheduling practice is dependent on a number of factors such as the design of the shop floor, the degree of complexity of the operations and the overall predictability of the manufacturing process. It is recognized that a well designed, simply organized and stable manufacturing process is easier to schedule than a more complex and volatile system. This point will be discussed in more detail below.

Much work has been reported on production scheduling (Graves, 1981), highlighting the fact that scheduling is a complex task, the technical difficulty being that of combinatorial explosiveness (Rickel, 1988). This magnitude of possibilities makes the goal of schedule optimality an unattainable ideal, and according to Baker (1974), there are 'relatively few situations in which general optimal solutions are known'. Regarding the ideal of optimality, it is important to note that 'the optimal solution of a model is *not* an optimal solution of a problem unless the model is a perfect representation of the problem' (Ackoff, 1977).

PUC is a *self-contained task* which controls a specific PU within the factory, and the scheduling function of PUC takes as its primary stimulus the schedule guidelines from the FC system. These guidelines specify the time constraints within which a series of job orders are to be completed, and the role of the PUC scheduler is to take these guidelines and develop a plan which can then be released to the shop floor via the PUC dispatcher. The actual development of the schedule may be based on any one of a number of techniques, algorithms or computer simulation packages.

The scheduling function includes three activities which are carried out in order to develop a realistic schedule for the shop floor.

- Firstly, a check on the system capacity is required, the objective being to calculate whether or not the schedule guidelines specified by the

GFC system are realistic. The method of doing the capacity analysis depends on the type of manufacturing environment, and the results of the capacity analysis will have two possible outcomes: either the guidelines are feasible or they are not. If the guidelines are feasible, the capacity constraints are then included in the procedure for developing a schedule.
- Secondly, a schedule must be generated. If there is a major problem with the available capacity, the scheduler may need to inform the GFC system, and the overall guidelines for scheduling that particular cell may have to be modified.
- Finally, the schedule is released to the dispatcher so that it may be implemented on the shop floor. In a CIM environment, this release will be achieved by means of a distributed software system, which will pass the schedule between the scheduling and dispatching functions.

Scheduling represents one aspect of PUC, that is, the planning aspect. The schedule is developed taking different constraints and variables into account. When it is released to the shop floor it becomes susceptible to the reality of shop floor activity, and in particular, the *unexpected event*. The unexpected event is a true test of any system's flexibility and adaptability, and it is the role of the dispatcher to deal with inevitable unplanned occurrences which threaten to disrupt the proposed schedule.

The dispatcher

Events such as machine breakdown, or quality problems, can have a serious effect on the production plans supplied by the scheduler. The main purpose of the dispatcher is to react to the current state of the production environment, and select the best possible course of action. In order to function correctly, the dispatcher requires the following important information:

1. The schedule, which details the timing of the different operations to be performed.
2. The state-independent manufacturing data describing how tasks are to be performed.
3. The current shop floor status.

Thus, access to the latest shop floor information is essential, so that the dispatcher can perform intelligent and informed decision making. In fact, one of the greatest obstacles to effective shop floor control is the lack of accurate and timely data. As Long (1984) pointed out, the possibility of a good decision is directly related to the integrity of the manufacturing data. The dispatcher may use different algorithms and procedures to ensure that the schedule is followed in the most effective way. When a decision has been made as to the next step to be taken in the production process, the dispatcher will send instructions to the mover and the producer so that these steps are carried out.

The three main activities of the dispatcher involve *receiving information, analysing alternatives and broadcasting the decision*. The Information received is the scheduling information, as well as both static and dynamic manufacturing data. The static data may be obtained from the manufacturing database while the dynamic information describing the current status of the shop floor is received from the monitor. When received, this data is collated and manipulated into a format suitable for analysis. This analysis may be carried out using a range of software tools, or performed manually by a supervisor, based on experience and intuition. The analysis will most likely take place keeping the overall dispatching goals in mind, and the end effect of this analysis is to broadcast an instruction to the relevant building block, perhaps using a distributed software system.

The implementation of a dispatcher varies depending on the technological and manufacturing constraints of a production system. The dispatching task may be carried out manually, semi-automatically or automatically. Manual dispatching involves a human decision on what the next task should be in the system. Examples of this might be an operator deciding to select a job according to some preference. This preference might be generated using a *dispatching rule*, which prioritizes jobs in work queues according to a particular parameter (e.g. earliest due date, or shortest processing time). A semi-automatic dispatcher may be a computerised application which selects jobs, but this selection can be modified by an operator. An automatic dispatcher is used in an automated environment, and it assumes responsibility for controlling the flow of jobs through the system.

To summarize, the dispatcher is the controlling element of the PUC architecture, and it ensures that the schedule is adhered to as far as possible. It works in real-time by receiving information from the monitor on the current state of the system and it issues instructions to the moving and producing devices so that the required work is performed.

The monitor
Within the different levels of manufacturing, from strategic planning down to PUC, informed and accurate decision making relies on consistent, precise and timely information. Within PUC, the monitor function supplies the necessary information to the scheduler and dispatcher, so that they can carry out their respective tasks of planning and control. Thus, the role of the monitor is to make sense of the multitude of data emanating from the shop floor, and to organise that data into concise, relevant and understandable information for the scheduler and dispatcher. Put simply, the monitor can be seen as a translator of *data* into *information*, for the purpose of providing sensible decision support for the scheduling and dispatching functions.

Higgins (1988) identified three main activities of the monitor as: *data capture, data analysis* and *decision support*. The *data capture* system collects data from the shop floor. This is then translated into information by the *data analysis* system, and can then be used as *decision support* for appropriate PUC activities.

Data Capture A vital part of the monitor is the data capture system, which makes the manufacturing data available in an accurate and timely format. This data capture function should perform reliably, quickly and accurately without detracting from the normal day-to-day tasks which are carried out by humans and machines on the shop floor. Ideally, data capture should be in real time with real time updating, and the data should be collected at source. Automatic or semi-automatic collection of data is often necessary for reasons of accuracy and speed of collection.

Data captured which may eventually be used for informed decision making at a higher level in the PUC architecture includes: process times, job and part status, inspection data, failure data, rework data, and workstation data.

Work in Progress Status	Job number
	Part name
	Current location
	Current operation
	Due date
	Number of remaining operations
Workstation status	Workstation name
	Current status
	Current job number
	Utilization
	Percentage time in set-up
	Percentage time processing
	Percentage time down

Table 30.1 *Typical information from the data analysis of the production monitor*

Data Analysis The data analysis function of the monitor seeks to understand the data emanating from the data capture system. It is a very important component of the monitor, because it takes time and effort to filter important information from a large quantity of shop floor data. Thus, this data analysis function effectively divides the monitor into different sub-monitors, which then keep track of different important aspects of the manufacturing system. Joyce (1986) identified three main classes of monitor:

1. Production monitor
2. Materials monitor
3. Quality monitor

We will now discuss each of the main classes of monitor in turn:

Production monitor
The production monitor is responsible for monitoring work in progress status and resource status on the shop floor. Table 30.1 illustrates the type of information produced as a result of the data analysis performed

by the production monitor. At a glance, production personnel can see important manufacturing information regarding the progress of the schedule. This information can then be used as the basis for informed decision making. An important feature of the production monitor is the ability to recognize the point when the schedule is not implementable, and then request a new, more realistic schedule from the scheduler.

Raw materials monitor

The raw materials monitor keeps track of the consumption of basic raw materials at each workstation in the process. Table 30.2 shows the type of information generated by this particular monitor, and the main purpose of such a monitor is to ensure that there are no shortages in raw materials at a particular location. This is achieved by comparing current levels of a particular raw material with the recommended reorder point, and indicating when raw materials need to be reordered.

Raw Material Status	Material name
	Workstation name
	Buffer name
	Current quality
	Reorder point
	Rate of usage

Table 30.2 Information from the data analysis of the materials monitor

Quality Monitor

As the name suggests, the quality monitor is concerned with quality related data, and aims to detect any potential problems. Quality problems may arise from internal or external sources. Quality problems arising from external sources originate in the supply of raw material purchased from vendors which can cause problems in later stages in the production process. Problems originating in the production process which affect the quality of products are classified as internal sources. These types of problems can be indicated by the yield of the cell or of a process within the cell. If the yield falls below a defined level, an investigation into the cause may be warranted. Possible causes of a fall in yield status can range from faulty equipment to poor quality raw materials.

Decision Support The main function of the monitor's decision support element is to provide intelligent advice and information to both the scheduling and dispatching functions within PUC in quasi-real time.

The mover
The mover coordinates the material handling function and interfaces between the dispatcher and the physical transportation and storage mechanisms on the shop floor. It supervises the progress of batches through a sequence of individual transportation steps. The physical realization of a mover depends on the type of manufacturing environment. It can range from an Automatically Guided Vehicle system (AGV) to a simple hand-operated trolley. The selection of the items to be moved is predetermined by the dispatcher and this decision is then transferred as a command to the mover, which then carries out the instruction. The mover monitors the states of the transporter and the storage points, translates the commands from the dispatcher to specifically selected moving devices, and also issues messages to the dispatcher signalling commencement and completion of an operation. An automated mover system might use collision avoidance algorithms, to ensure that no individual device will cross the path of another when parts are being moved to their destinations.

The producer
The producer is the process control system within PUC which contains (or has access to) all the information required to execute the various operations at that workstation. The producer may be an automated function or a human. The main stimulus for a producer comes in the form of specific instructions from the dispatcher building block, and these instructions specify which batch to process. The producer accesses the relevant part programs (detailed instructions of the operation that has to be performed), and also the configuration data which specifies the necessary set-up steps that are needed before an operation can commence. The producer translates the data into specific device instructions and informs the monitor when certain stages of activity have been completed (e.g. set-up completed, job started, job finished, producer failed, etc.).

REFERENCES

1. Graves, S.C., A Review of Production Scheduling, *Operations Research* 29 (4), 1981.
2. Rickel, J., *Issues in the Design of Scheduling Systems, Expert Systems and Intelligent Manufacturing*, M.D. Oliff (ed.), Elsevier Science Publishing Company, pp.70-89, 1988.
3. Baker, K.R., *Introduction to Sequencing and Scheduling*, John Wiley, New York, 1974.
4. Ackoff, R.L., Optimization+objectivity=opt out, *European Journal of Operational Research* 1, 1977.
5. Long, M.W., *Guidelines for the Collection and Use of Shop Floor Data, Execution and Control Systems*, Auerbach Publishers Inc., 1984.
6. Higgins, P.D. and J. Browne, The Monitor in Production Activity Control Systems, *Production Planning and Control* 1 (1), 1989.
7. Joyce, R., *Functional Specification of a Production Activity Control System in a CIM Environment*, Masters of Engineering Science Thesis, Industrial Engineering Department, University College Galway, Ireland, 1986.

Part F
MODELS OF THE FOF WORKBENCH

31

Introduction to the FOF Workbench

31.1 INTRODUCTION

The purpose of this part - models of the FOF Workbench - has been to provide the reader, a practitioner or a researcher, with some concrete tools to work with when dealing with customer driven manufacturing. The factory of the future is not the most tangible artefact in industrial society. It has been our intention to reduce the gap between the state-of-the-art and the future factory. This intention has here the shape of a bunch of conceptual and calculational models to approach and play with the concepts and relationships introduced and discussed in the previous sections of this book.

This set of models, called the FOF Workbench, is not intended to be an exhaustive toolbox to clarify all the aspects of customer driven manufacturing. It rather represents some vitally important facets of the problem domain. However, the conceptualization tool, REMBRANDT and the manufacturing systems design tools XBE are exceptions in this respect. They are generic and general tools to cover context dependently (REMBRANDT) and context independently (XBE-BET) the whole domain of customer driven manufacturing systems.

We shall introduce the following tools in the workbench:

- REMBRANDT-Reference model browser and design tool
- Human resources management tool
- Group design model
- Departmental coordination model
- Inter departmental coordination model
- XBE-BET - Expert System by Examples - Business engineering tool

REMBRANDT is intended to be a tool to provide an overview of all design choices and performance indicators with their interconnections. REM-

BRANDT thus serves as an intelligent dictionary for browsing through the concepts, relations and theories relevant for customer driven manufacturing. The relationships implemented are qualitative, indicating the dependency or impact of one system variable over another. Pairwise relationships have been implemented in the tool.

The human resources management tool is oriented towards strategic human resources management. It tackles the qualitative and quantitative resourcing problems, such as over/under resources, competence levels and aging profiles. Quantitative modelling of ill specified problems such as human resources management with simulation capabilities brings in-depth insight to the problem area and thus broadens and deepens the basis of decision making within organisations. It should be borne in mind that human resources management problems, e.g. shortage of capabilities for a new business environment, require long term proactive counter measures in education and recruiting. These counter measures are hard to identify and quantify without appropriate decision support tools, such as the one introduced in this connection.

The group design model is intended to help managers in the optimization of working group composition. It also helps to keep capacities and capabilities of the groups on the appropriate level to meet the business driven requirements. In other words, the group design model provides the operations and personnel managers with ideas on the organizational issues of group work. The design choices considered deal with the speciality of work allocation, rewarding, working conditions, training, personnel policy, communication, availability of extra capacity and recruitment. The performance measures under study deal with job depth and width and satisfaction, overtime, health, absenteeism and turnover. The underlying basic theories behind the reasoning in the model are related to methodologies invented by the Human Relations school of organizational theory, but also employed by engineers such as John Burbidge.

The departmental coordination model aims at illustrating how production control problems are influenced by the organisational structure. Thus the type of the primary process organization is the main design choice focused on. The relevant performance indicators in this model deal with costs, payoffs, controllability and lead time. The most fundamental factor under study is the decision making frequency. The departmental coordination model also relies on the philosophies of Burbidge.

The interdepartmental coordination model supports a manufacturing systems designer in developing an interdepartmental control structure when moving towards engineer-to-order mode from the make-to-order situation. The tool focuses on two particular design choices: the level of the control structure of customer order acceptance and the frequency of decision making at the aggregate production planning level. The most important part of this model is its relevance towards customer order acceptance.

XBE-BET covers both a software generator and an application. XBE-Expert System by Examples - is a generic tool for storing, manipulating and accessing expert knowledge via the case based reasoning methodology. In this domain a software tool called BET, business engineering tool, is a data "vacuum cleaner" and recycling device operating on the existing and past best practices of manufacturing systems design and operation. XBE-BET is definitely organized along the lines of the reasoning of senior consultants. It exploits the rational, theoretical as well as empirical, i.e. practical know-how and experience gained in the solution process of previous similar problems. Therefore the knowledge dissemination potential of the tool is interesting.

32

An intelligent storage and retrieval for design choices and performance indicators

32.1 INTRODUCTION

As is described in Chapter 2, "A design oriented approach", the Factory of the Future project has defined a modelling framework containing layers of models describing customer order driven manufacturing systems. The modelling framework is an open architecture in which knowledge from various disciplines can be interrelated for application to a particular problem.

The layers of the modelling framework all have different levels of abstraction. Also the upper layers are general models, applicable to various organisations, whereas the lower levels contain real data of a company in so-called particular models.

This chapter deals with the highest level of abstraction in the framework called the "conceptual reference model". Such a conceptual reference model provides a kind of qualitative connectance overview of design choices and performance indicators. This means the model shows all design choices (causes) and performance indicators (effects), related to the design and re-design of a customer order driven manufacturing system, with their interconnections. The interconnections are only described in a qualitative way (X affects Y, X increases Y, X simplifies Y, etc.).

The main goal of having an overview of all design choices and performance indicators with their interconnections is to quickly create insight into available theory. This insight consists of knowing the kind and power of existing relationships between redesign issues.

32.2 REMBRANDT

The idea of having a conceptual reference model containing all design choices and performance indicators with their interconnections seems very simple yet powerful. However, being able to work with such a model is something else. The amount of design choices and performance indicators exceeds very quickly the 200 mark and the relationships between these nodes are even a multiplication of that amount. It is clear that a paper representation of such a model is not usable and maintainable anymore. Creating insight to the model and being able to maintain it therefore requires computer support. For this purpose a tool has been developed by the FOF consortium called "REMBRANDT", an acronym for **RE**ference **M**odel **B**rowser **AN**d **D**esign **T**ool. REMBRANDT supports the following actions:

- browsing through the network of design choices and performance indicators (the DCPI network);
- creating insight to the type of relationship between these nodes[1];
- adding, changing or deleting design choices, performance indicators and their interconnections;
- "zooming in" on the quantitative aspects of (relationships between) nodes by enabling access to a particular model containing real-life date and simulation capabilities (see Chapters 33 to 37).

In the following subsections the requirements for using REMBRANDT are explained and thereafter the way of using it. The section on using REMBRANDT also goes into detail on the ideas and functionalities mentioned above.

32.3 SOFTWARE AND HARDWARE PLATFORM

REMBRANDT was developed for a Apple Macintosh platform, using the HyperCard 2.0 software package. HyperCard is a software tool which is delivered freely together with every Macintosh and therefore easily and widely available. The tool provides a quick prototyping environment with capabilities as an object-oriented user interface, hypertext,

---

1   A node is either a design choice or a performance indicator.

various real-time data inclusions from other applications, multiple and sizeable windows, etc.

32.4 USING REMBRANDT

The usage of REMBRANDT can be divided into browsing through the DCPI network and the maintenance of the nodes and relationships within the network. For covering this, REMBRANDT consists of three parts:

- the nodes and relationships database (browse and maintain)
- the graphical presentation of the network (browse only)
- the browser (browse only).

In the following the meaning and usage of these three parts will be discussed.

Nodes and relationships database
The database of REMBRANDT contains all design choices and performance indicators together with their relationships. These database contents are presented to the user as is shown in figure 32.1. In this figure we see the name of the node together with a long description. At the bottom of the screen, the connections are displayed. These connections are the relationships between the node and other nodes. In figure 32.1 we can e.g., see that the node *product quality* is affected by *skill range*. Of course when we take a look at *skill range* we will find the opposite relationship: *skill range* affects *product quality*.
In working with the node and relationship database, the user is able to search for a particular node and look into the definition and relationships of that particular node. However, he has not only the opportunity to browse through but also to maintain the database. This is done by adding new nodes or by changing or deleting existing ones. When adding or changing nodes, one can define new relationships or delete relationships with other existing nodes.

The types of connection which can be defined between two nodes are listed in table 32.1.

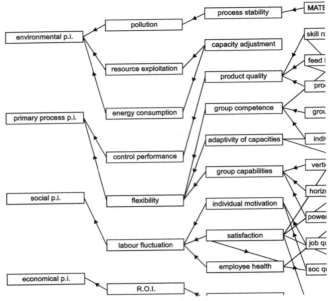

Fig. 32.1 Example of a database representation screen

Affects	Affected by
Increases	Increased by
Decreases	Decreased by
Improves	Improved by
Worsens	Worsened by
Simplifies	Simplified by
Complicates	Complicated by
Benefit of	Has benefit
Element of	Has element
Pertains to	Pertaining to

Table 32.1 Connection types between nodes

As table 32.1 shows, there are only qualitative relationships defined between nodes. There is, however, a possibility to say a little bit more about the power of a relationship. This is done by stating per relationship if it is strong, normal or weak. It is obvious that using these terms is very difficult and, even if there are strict rules in choosing one of the terms, a subjective point.

32.5 GRAPHICAL PRESENTATION OF THE NETWORK

Of course browsing through a text oriented database does not give much insight to the DCPI network. Because of this a graphical network browser was created (see fig. 32.2, which is identical to fig. 2.2). This graphical browser can display all nodes with all relationships. However, because of the large amount of nodes and relationships, even this is too complex to understand for human beings. Therefore the graphical browser enables the creation of sub-networks.

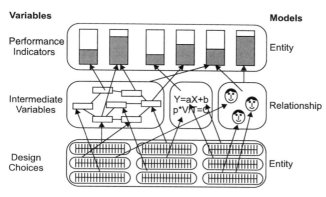

Fig. 32.2 Part of a graphical representation of the DCPI-network

Creating a sub-network is done in the following way:

- select one or more nodes as being the nodes to start with;
- select one or more relationship types of table 32.1 to follow from the starting nodes. This results in an initial sub-network.

In the sub-network for example the starting node employee satisfaction is chosen and the type of relationship is affected by. REMBRANDT now starts looking for the nodes which affect employee satisfaction directly. These nodes are than added to the network and a line is drawn between the added node and the start node to represent the relationship. Now the tool searches further from the added nodes to see what they are affected by, then these nodes and relationships are added as well, although influencing employee *satisfaction indirectly*. Again from the

last added nodes the search is continued, nodes and relationships added, and so on.
- by choosing one or more nodes and relationship types the user does not know how the resulting sub-network will look in advance. This means that the user generally in the first step ends up with either a too simple or a too complex sub-network. It may be too complex because of the following reasons:
  - there are too many starting nodes
  - the starting nodes are very important design choices or performance indicators for a company (like *flexibility* or *total costs*) and act therefore as the pivot on which everything hinges
  - the relationship type(s) enclose too much. When e.g. the *affects* relationship is chosen, automatically *improves, decreases, worsens,* etc. are taken into account.

On the other hand it may be too simple because of the opposite reasons: too little, unimportant starting nodes combined with too narrow relationship types.

For all these reasons, drawing a sub-network will be an iterative process in which you gain more knowledge and insight at every step.
- If a sub-network is drawn which offers sufficient overview and insight the user can do the following:
  - zoom in on a node of the network to get more information (description, other not graphically shown relationships, etc.) directly from the database
  - analyse a relationship by zooming in on it. This results in the relationship type and power
  - choosing one of the particular models (see Chapters 33-37) related to the selected node, to perform quantitative calculations and simulations on a real-life situation.

32.6 CONCLUSION

This chapter has shown, how the ideas presented in Chapter 2 can be implemented in a computerised tool. In a practical case, Design Choices and Performance Indicators should be defined for a specific context, and with specific numerical values. However, the previous parts of this book have shown that it is also possible to relate design choices and performance indicators in a more general way, according to well established (scientific) knowledge. The description of REMBRANDT serves as a vehicle to underscore that this knowledge is explicitly available.

33
A simulation model for human resource management in customer driven manufacturing

33.1 MOTIVATION

Human Resource Management is often simply considered to provide the capacities derived from the production programme. For this purpose capacity requirements will be predicted in terms of required man hours or man months for different functions like welding, painting or assembly work. Based on a given product or production programme, the number of man hours required at a given time for welding, painting and assembly work are calculated.

The available capacity is then usually calculated by the number of employees multiplied by their daily working hours. This results in a quantitative measure for the available capacity. In this approach, Human Resource Management is the task of hiring or firing employees on time or negotiating with the staff on overtime in order to match required and available capacities.

In Tayloristic production systems, where human work is reduced to a functional element, this approach might be adequate. But in many other cases human work requires more than to be hired. For companies relying on living knowledge and experience, it is essential to provide for the internal transfer of knowledge and experience gained from production (see Chapter 17).

For these companies it is very important to consider not only the number of employees in the company but also their age distribution. This chapter will show that age distribution is a simple, but informative method for discovering the need for Human Resource Management.

408   A simulation model for human resource management

To illustrate the importance of age distribution, the following constructed example shows the same total number of employees in four different age distributions.

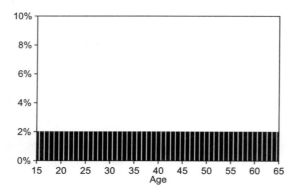

Fig. 33.1   Equal age distribution of employees

In the first case (figure 33.1) an equal distribution is assumed. An equal distribution of employees aged from 15 to 65 is unlikely in reality, but the calculation on total figures at least implicitly assumes such a distribution.

In the second picture (figure 33.2) the need for dedicated Human Resource Management becomes clearer. This distribution shows a peak in the 20-25 age group, another peak around 54 and a depression in the 38-50 age group. If we now assume that the employees around 54 are the most experienced in that company and, moreover, if we assume that the internal experience transfer takes up to ten years (Hamacher, 1989), it becomes evident that this company is going to be faced with the critical situation of their most experienced employees retiring within the next ten years. The age distribution indicates an urgent need for immediate action to support the transfer of experience to the younger employees, otherwise significant company experience will be lost.

This age distribution also shows that this transfer needs to be managed in a situation where the coherence and mutual understanding between the employees is probably unbalanced. The two peaks in the distribution also represent a segregation into two different generations. A clear segregation of this kind is usually an indicator for cultural differences and low mutual understanding between age groups, which will not facilitate a process of experience transfer. To summarise, figure

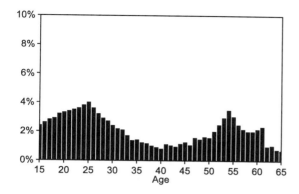

Fig. 33.2 Bi-modal age distribution of employees

33.2 represents a situation in which the management has failed to initiate appropriate Human Resource Management action in time.

The third distribution (figure 33.3) represents a situation where the opportunity to retain the living knowledge and experience of the company has already passed. The employees are very homogeneous in age, but very old. If we assume that a high age correlates with "a long time together", the coherence and mutual understanding within this staff can be expected to be well-developed. But in this case an internal experience transfer will only be successful if transfer times are short. In all other instances the experts will have disappeared before the transfer has been completed. This situation may be welcome if a restructuring process is envisaged, but experience and relationships -

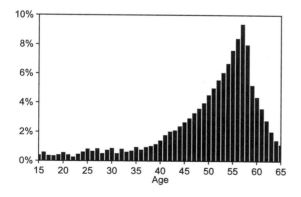

Fig. 33.3 Right-peak age distribution of employees

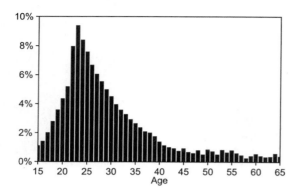

Fig. 33.4 Left-peak age distribution of employees

including the personal links to customers and suppliers - will have to be rebuilt from scratch.

If, finally, we look at the fourth distribution (figure 33.4), we see a homogeneous young staff. From a distribution of this kind we can expect developed mutual understanding, a specific coherence, but low experience. This, for example, could represent the situation in a software house: a high level of motivation and readiness for innovation, but still low experience in real business needs.

This constructed example should show that an analysis of the age distribution can provide insights into the future availability of human resources and respective means for influencing this living development towards the company's objectives. But a dynamic analysis of such age distributions requires a lot of experience and knowledge of the "mechanics" of social groups and the theories on which it is based.

This was another reason for covering existing theories on learning, human behaviour and social groups in a simulation model which can support a dynamic analysis of age distribution in a company.

33.2 BASIC APPROACH

The basic approach and the main assumptions for the model will be outlined below. For the design of a simulation model it is assumed that the abilities of an employee will change during his working life (Lutz, 1988). At the start of working life an employee usually has low experience, but a strong desire to change structures and processes.

This innovativeness decreases over time, when employees accommodate themselves to existing structures and processes.

The growth of experience is basically considered as a function of time as well (Dreyfuss, 1986), but it is also influenced by the context of work. If in the work context many experienced colleagues are available and time for communication and exchange is procured, the growth process of experience will be forced. If, on the other hand, there are only beginners and there is no time to communicate, the growth of experience will be diminished.

To derive empirical measures from the age distribution of a specific company it is assumed that the age distribution of the employees in a company correlates with the time they have worked in that company or at least with their professional life. Starting with an entry age of 15, this means that an age of 55 is interpreted as 40 years of work in a company or profession. This interpretation of an age distribution is, of course, an oversimplification. A simple age distribution does not cover the effects of fluctuation or the fact that in many cases professional life starts at an age of 25 or 30. For the purpose intended, this will be disregarded here.

Bearing these limitations in mind, a given age distribution can be translated into a set of quantitative performance indicators (recall that the concept of performance indicators was discussed in Chapter 2). In this model in particular, four different performance indicators are considered:

1. work capacity
2. experience
3. innovativeness
4. coherence

Work capacity is measured in man-hours per year and gives an indication of the global capacity of a given workforce. This indicator is calculated by multiplying the number of employees by their average annual working time in hours.

The experience of a given workforce is calculated on the basis of the time (in years) they have been in work. This indicator is calculated as the total sum produced by multiplying the number of employees in each age category by the number of years they have been in work. As mentioned above, this calculation is age minus 15. As the increment of experience is not linear, but decreases over time, the performance indicator is calculated as the square root of the total sum.

412 A simulation model for human resource management

For innovativeness it is assumed that each employee entering the company starts with a level of 100, which will decrease as a logarithmic function down to 50 over 5 years. The performance indicator is the total sum of these figures.

Coherence also refers to the number of years people have been working together. So the total sum of working years is again considered as one factor to calculate a performance indicator for coherence. But unlike coherence, two other factors are considered: one is the ratio of turnover to the total number of employees. This ratio considers that a large turnover in personnel will spoil the level of coherence. The second factor is modelled as the time fraction of the total average working hours available for interaction. This factor considers that time available for interaction will speed up the development of coherence within the work force. So the performance indicator for coherence is ultimately built up from measures of working years, turnover and interaction time.

Performance indicators by definition are empirical measures for objectives. They are means for monitoring a process, but not for influencing the process to be monitored. But a good simulation model should provide some means for influencing the process in order to provide insights into actions and their outcomes.

The model created provides two kinds of action, which are called design choices (cf. Chapter 2). These design choices relate to

1. hiring strategy
2. the volume and structure of working hours.

The hiring strategy implemented in the model allows a target value for the number of employees to be selected and a replacement factor for the employees leaving the company due to retirement to be set. The model provides a governor for how many employees will be hired to replace employees who have retired. The retirement process is considered as not being influenced. Employees above the age of 65 will be automatically pensioned off. No other "firing" procedures are implemented.

The hiring factor influences overall human capacity, the level of experience, the level of innovativeness and the coherence within a workforce. The second set of design choices implemented in the model allows the total working hours and the fraction of working hours available for indirect work like interaction, exchange of experience and training to be changed. The impact of these "governors" on capacity adjustments is easy and obvious. The influence on the other three performance indicators requires more complex considerations. Although some theories are available to model the impact of working hours on innovativeness and experience, up to now the relations have only been modelled for indicator coherence. Here it is assumed that an increased fraction of working hours for communication will raise the coherence level. This is again modelled as a square-root function.

To create a simulation model for Human Resource Management a "Systems Dynamics" approach (Forrester, 1961) was initially selected using the known concepts of "levels", "rates" and "converters" to define simulation models as an interrelated, time-dependent set of equations. As Systems Dynamics-type simulation usually cannot consider attribute changes at an individual level, a revised model was subsequently developed for a "microsimulation" approach (Brennecke, 1975).

33.3 TECHNICAL STRUCTURE OF THE SIMULATION MODEL

For the development of the Systems Dynamics model, STELLA was used on a MacIntosh as a hardware/software platform. STELLA is a very powerful software environment for developing and testing Systems Dynamics models and was used in this way. To ease the use of the model, a user-interface written in HYPERTALK was developed and linked to the STELLA model. This user-interface provides four different sections for the user: one input section to edit the age distribution and other characteristics of a real company, a second section to preset the design choices, a third section to start the simulation run and a fourth section to review the output of a simulation run by a structured set of diagrams.

The reimplementation on a microsimulation approach was done in an EXCEL 4.0 environment. This model consists of a database section, an "Event Generator" to manipulate the database according to the modelled design choices and an "Evaluator" to generate the statistics and diagrams.

Specimen examples

To illustrate the use of the Human Resource Management simulation model, a simple example using this model will be given below.

This example uses the age distribution shown in figure 33.2 as the starting point for further simulation runs. The diagram shows that 11.8% of the actual staff will leave the company within the next ten years

414 A simulation model for human resource management

through retirement. This is equivalent to a reduction of 11.8% of the human resources if no action is taken. Furthermore, the diagram indicates that the company will lose most of its experienced employees within these ten years. So let us examine what the probable effects will be if we apply two alternative strategies.

The first strategy selected, called "capacity replacement", is to hire a young employee from the labour market immediately for each retiring employee in order to maintain the initial capacity. The results of the simulation run over ten years is shown in figure 33.5.

The diagram shows that the capacity indicator remains constant. This is intended and not astonishing and merely shows proper calculation. More interesting, and not so obvious, is the behaviour of the experience indicator. In the first three years the loss of experienced employees will be overcompensated by the initial growth of experience in the young employees. There are also still sufficient "seniors" available to maintain the transfer process. But after three years the growth converts into a rapid decline in the level of experience: among the majority of "seniors" the context and mechanism for a proper transfer process has gone - the experience embodied in the company diminishes.

With regard to innovativeness the diagram shows that hiring "fresh" and innovative employees will not compensate for the overall loss of innovativeness. To maintain the level of innovativeness other action - like training - is required, affecting existing staff as well.

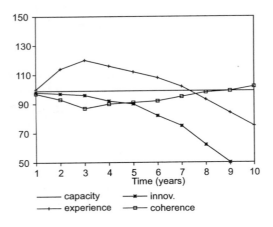

Fig. 33.5 Results of "capacity replacement"

The fourth indicator, coherence, shows an initial decline and thereafter a steady growth in coherence. The explanation for this process is simple: in the beginning the disruption caused by changing personnel results in a decline, but at the end the homogeneity of the staff is higher than at the beginning as most of the employees are young. This in turn leads to a rise in this indicator's score.

To summarise, the "capacity replacement" strategy shows some unexpected side effects, which remain hidden to managers as a result of merely calculating capacities. But let us look now at what the second strategy, called "experience transfer", will produce.

This strategy focuses on the maintenance of existing experience. To achieve this, 10% of the working hours will be reserved for interaction and exchange within the existing staff. The result is a total capacity drop of 21.8%. However, the positive effect of this action is that the level of experience immediately rises, but almost levelling out after five years once the existing experience has been absorbed.

The other positive effect of this strategy is a significant increase in coherence. This increase is even higher than the increase in experience and thus proves the power of communication to establish social relations.

However, this strategy does not help to increase innovativeness. The respective indicator is cut within eight years by half. Looking at this more closely, it is understandable, as more time to communicate alone cannot result in innovation. To achieve this, it will be necessary to feed in new knowledge from outside.

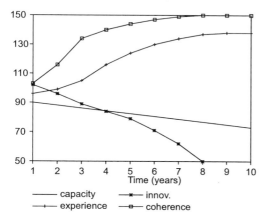

Fig. 33.6 Results of "experience transfer"

Summarising both examples, the first case may be typical for enterprises providing only for capacity maintenance: the loss of experience here is often offset by predefined processes and procedures in order to become independent of living experience. The second case also has a frequent counterpart in reality: instead of applying smooth and balanced action, massive responses will be applied to push one indicator in the desired direction. This may solve the experience-transfer problem, but the sequence of peaks and troughs in the distribution will be extended to another cycle.

However, the most important lesson is that complex problems cannot be solved by single and isolated measures, they usually require a carefully selected mix. Simulation models like the model presented here may help to understand the mechanics of human activity systems and can provide some initial guidelines, but social reality is always much more complex. So, simulation may be a valuable tool, but is definitely no alternative to reality.

REFERENCES

1. Brennecke R., *Die Konstruktion von sozioökonomischen Groásystemen*, Frankfurt/New York, 1975
2. Dreyfuss H.& S., *Mind over Machine*, Blackwell 1986
3. Forrester J.W., *Industrial Dynamics*, Cambridge Mass., 1961
4. Luhmann N., *Soziale Systeme - Grundriá einer allgemeinen Theorie* -Frankfurt/M 1987
5. Lutz B., *Qualifizierte Gruppenarbeit -überlegungen zu einem Orientierungskonzept technisch-organisatorischer Gestaltung*, Karlsruhe 1988

STELLA is a trademark of High Performance Inc. USA

34

Group Design

34.1 INTRODUCTION

In customer driven manufacturing, the availability of the right kind and quantity of resources able to engineer, manufacture and assemble a product in line with the customer's needs is very important. The term "resources" refers to machines, tools, transports, people, etc. As has been emphasised in part A and in Chapter 17 of part B, the focus in customer driven manufacturing should be on human resources rather than physical resources. This is because human resources, which provide creativity, experience and awareness, are currently seen as the most driving force in OKP systems.

Human resources can be considered over a long period of time, balancing factors like innovativeness, skills, experience and the coherence of and between people (see Chapter 33). In balancing the factors mentioned over a long period of time, the point of interest is not how human resources are divided into groups over different departments but how the total of human resources behave over time. When looking at human resources over shorter periods of time, the distribution of people over skills and groups becomes important. In customer driven manufacturing systems especially, autonomous and qualified working groups are the basic building blocks for meeting all the required work. These working groups can be seen as the driving forces in finding and implementing the most appropriate solutions to the customer's special needs.

Modelling human capabilities, combined into small working groups, over a short and medium term period is the basis on which the Group Design model has been created. The Group Design model tries to answer the question of what managers can do to obtain the best possible composition of working groups, keeping the capacities and capabilities of these groups at the appropriate level to meet all the required work. This means that the Group Design model tries to

418 *Group Design*

provide the operations manager or the personnel manager, for example, with some solutions, or at least ideas, to keep their groups organised in the right manner.

To get an answer to the question addressed by the Group Design model the following is required:

- a method for describing the capacities and capabilities of working groups
- design choices to influence these capacities and capabilities
- performance indicators showing how good or bad a group performs on several production and social factors
- a way to relate the three issues above: a model describing working groups together with possible actions to influence the capacities and capabilities of these groups and the way these actions result in changes in the group's performance.

The above points of the Group Design model will be discussed separately.

34.2 DESCRIBING CAPACITIES AND CAPABILITIES OF WORKING GROUPS

The autonomous and qualified working groups, able to produce complex one-of-a-kind products in an active and cooperative manner, are part of the organisation's network of working groups. Each group typically consists of six to twelve people. The capacities and capabilities of each member of a working group need to be described individually. This is done using a matrix (see figure 34.1).

The job functions a person can execute (drilling, welding, etc.) are listed on the horizontal axis of this matrix. On the vertical axis we see four levels at which each job function can be performed: the work preparation level, the work planning level, the execution level and the quality control level (feedback on how well the job has been done). We call these levels the job levels. This means that one cell of the matrix defines a job function performed at one job level. This cell is called a *capability*. By completing the matrix shown in figure 34.1 the number of man hours available or required per time period can be listed in every cell. These hours define the *capacity* for every capability listed in the matrix.

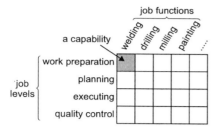

Fig. 34.1 The capacities/capabilities matrix (the c/c matrix)

For example: if the dark grey cell in figure 34-1 contains the number 30, the capacity desired or available (depending on the context) for preparing the work for drilling jobs is 30 man-hours. Because the matrix in figure 34.1 describes the capacities per capability, we shall refer to this matrix from now on as the *capabilities/capacities* matrix (the c/c matrix).

The c/c matrix can be used at different levels of abstraction. It can be filled for just one person, for a whole group or even for a department or whole organisation. Since the Group Design model deals with groups, the individual matrices of the group members are added into one group matrix and the operations are performed as much as possible on this group matrix and not on the individual matrices.

As mentioned above, the c/c matrix can both be used to describe the required work (capacities per capability) and to describe the work really done by the group. Suppose we fill in the c/c matrix of a group over a certain period P for the work this group had to perform (the *required* c/c matrix). Another matrix is filled in showing the work actually done by the group during period P (the *allocated* c/c matrix). Then, the difference between these two matrices gives us an indication of how well or badly the group was organised doing the work within period P (see figure 34.2). By organised is meant the grouping of the employees so an optimum combination of capacities and capabilities can be achieved. The difference between the two matrices (which will be called DELTA (Δ) from now on) will then be a criterion by which to decide to change certain group characteristics.

Stating that the required c/c matrix and the allocated c/c matrix form the bases on which to take action, we first need to check how these matrices are created. The required c/c matrix simply shows the work given to the group and can therefore be derived direct from the work flow. The allocated c/c matrix, showing the work actually done by the

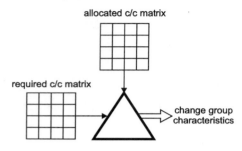

Fig. 34.2 Required versus allocated c/c matrix

group, depends on the work given to the group and the capacities and capabilities of the group.

The capabilities of the group members will be described through productivity factors. These productivity factors have a range from zero to one. "Zero" means that an employee is unable to perform this capability; "one" means the employee is a specialist in performing this capability. The productivity factors of an employee can again be presented as a c/c matrix. The only difference is that the values of the matrix are not measured in man-hours but in numbers between zero and one.

When we multiply the productivity factors matrix of each individual group member by the number of working hours, we create another matrix showing the maximum available man-hours for every capability (the maximum available c/c matrix) per employee. By matching the maximum available c/c matrices, the required matrix and the number of working hours per period, we are able to create an allocated c/c matrix (see figure 34.3).

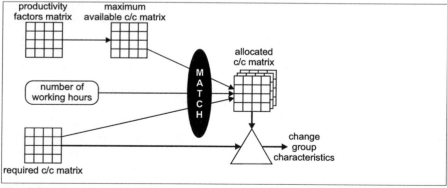

Fig. 34.3 Creating the allocated c/c matrix

34.3 DESIGN CHOICES

Determining the design choices to influence the capacities and capabilities when DELTA becomes too large is the next step in creating the Group Design model. The possible decisions to change the group characteristics are called design choices (DCs). We distinguish seven DCs in the Group Design model:

1. *Specialist versus generalist work allocation*
   There are two ways of composing groups. The first is to work with specialist employees who are very good at only a few capabilities. The second is to have employees who can work on many capabilities but are not as good in a single capability as a specialist. These employees are called generalists. The implications of the kind of employees chosen may be great. Specialists, for example, can do more work on a few capabilities but may be dissatisfied with their work because it may become dull. Generalists on the other hand may be satisfied with their work because they work on a wide variety of capabilities, but may not be fast or good enough for the jobs to be done. By setting this DC, a manager can decide how to achieve the match to arrive at an allocated c/c matrix (see figure 34.3). The way these allocation methods work is described in the next section, which concerns the implementation of the Group Design model.
2. *Rewards*
   The rewards (remuneration) given to the employees may have implications for the satisfaction of these employees and therefore for their work.
3. *Working conditions*
   Working conditions can have a big influence on the satisfaction and health of the employees. Therefore, changing the working conditions may have implications for the group's performance.
4. *Training*
   Training employees in certain capabilities leads to an increase in the available capacities and/or capabilities of the group.
5. *Personnel policy*
   Personnel policy is the company's policy focusing on targets in terms of the quality and quantity of present and future human resources. An employee friendly personnel policy will keep employees more satisfied than an unfriendly one.
6. *Communication*
   The time allowed for communication between employees (formal

and informal) can have an impact on the satisfaction of the employees and therefore on the work done by these employees.
7. *Selection*
The selection and employment of new workers for a group has an obvious influence on the availability of capacities per capability.
8. *Maximum overtime*
The maximum number of hours overtime allowed over a certain period can affect the satisfaction of the employees and the availability of capabilities and is therefore the final DC of the Group Design model.

34.4 PERFORMANCE INDICATORS

All the DCs mentioned above have implications for factors concerned with group design. To see if the decisions made in designing working groups (the settings of the DCs) were the right ones, another kind of variable can be looked at: performance indicators (PIs). The PIs of the Group Design model are the following:

1. *Job depth*
Job depth is an indicator for the job levels (work preparation, planning, execution, quality control) at which the capabilities of an employee are performed. When an employee has to work at every job level, his job depth is high (the job is "deep"); when he only works at one job level his job depth is low.
2. *Job width*
Job width is an indicator for the number of job functions (drilling, milling, etc.) worked on at every job level. When an employee works on many job functions, his job width is high; when he works on only a few his job width is low.
3. *Overtime*
The number of hours overtime worked by the employees. (This is different from the "maximum overtime" DC because the DC only gives an upper limit to this PI: the overtime actually worked).
4. *Satisfaction*
The employees' job satisfaction.
5. *Health*
The average health of the group members.
6. *Absenteeism*
Absenteeism is an indicator for the amount of time employees fail

to show up at work (when they should have). There may be several reasons for this: illness, dissatisfaction, etc.

7. *Turnover*

Turnover is a rate giving an indication of the number of employees leaving the company. One reason for leaving the company could be the normal switching from one company to another because of better remuneration, etc., but an employee's departure could also be due to dissatisfaction with his or her current work.

8. *DELTA*

As mentioned above, DELTA is the difference between the work to be done by the group and the work actually done by the group. This means that when the average DELTA is high, there is a big gap between required and allocated work and so the group is badly designed or receives badly allocated work. When we say that DELTA is high, we either mean that the group receives much more work than it can handle or it receives too little work so the workers almost have nothing to do.

34.5 RELATING DESIGN CHOICES AND PERFORMANCE INDICATORS

The entire Group Design model is displayed in figure 34.4. The DCs of the model are marked with thick lines. In the illustration we can see some DCs and PIs which have more than one level. This means that these DCs and PIs have to be calculated at an individual level (but they can be presented to the user of the model by an average group level).

Each arrow in figure 34.4 represents a relationship between two variables (design choices, performance indicators or intermediate variables). In making a simulation tool for the Group Design model, these relationships can sometimes be described by simple formulas (e.g. for the relationship between absenteeism and working hours), sometimes by heuristic algorithm (e.g. matching) and in all other cases by a graphical representation of the relationship (e.g. the relationship between satisfaction and rewards or between satisfaction and working conditions; see figure 34.5). The relationships mentioned are found in the literature.

Finally, a short overall description can be given of how managers will actually work with the Group Design model or a simulation tool of this model. Firstly, an existing or desired situation is defined. This means a description has to be created of one or more working groups,

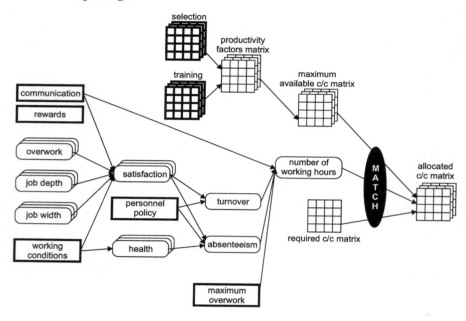

Fig. 34.4 The Group Design model

together with a description of the capacities and capabilities of the people involved (using the appropriate matrices). Secondly, the design choices have to be set to an initial value. After that, the performance indicators can be determined by the model. By analysing the performance indicators, it is possible to decide to change the design choices and therefore the characteristics of the working group. Again, the model can determine the changed performance indicators. In this way a manager can see the implications of his actions and go through an iterative process of improving the design of working groups.

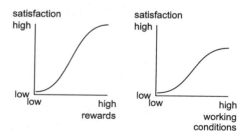

Fig. 34.5 Examples of graphical relationships in the Group Design model

34.6 CONCLUSION

This chapter has shown how the ideas presented in Chapter 17 can be operationalised for a particular company, department or group. Whereas the REMBRANDT tool presented in Chapter 32 remains at a general, qualitative level of reasoning, the models presented in Chapters 33 and 34 can be loaded with actual company data. This is a useful tool for evaluating a company's situation and for a systematic discussion of alternatives which lead to improvement.

35

Departmental coordination model

35.1 BUSINESS PROBLEMS ADDRESSED BY THE MODEL

The departmental coordination model aims at illustrating how the production control problem is influenced by the organisation of the primary process. The organisation of the primary process can move from a functional organisation to a product-oriented organisation (Group Technology organisation, see Chapter 15). Depending on the type of organisation, the performance of the primary process is different. Also, the complexity of the control will depend on this organisation of the primary process. Thus, the primary process ("type of process") organisation is the main Design Choice.

The model has been developed at the level of a production department. The following picture illustrates the domain studied through this model (figure 35.1).

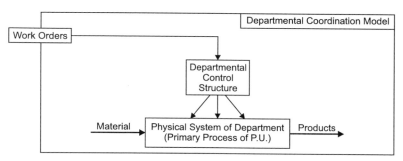

Fig. 35.1 Domain of the study

In repetitive production, the designers of production systems often focus on flow production: optimisation and simplification of flows is the control issue. The Group Technology (GT) approach provides a good tool for that purpose. In flow production, the management of the flows (products, information,...) are less complex and the production cycles are shorter. However, GT supposes a minimum of standardisation, or at least an identification of stable product families. Such a situation leads to less flexibility in terms of product types, and in terms of capability of adaptation to various customer needs.

In this departmental coordination model, several assumptions have been made. All the implemented relationships and the results of the simulations are depending on these assumptions. They are described in the last section of this chapter.

This model does not claim to design a perfect control structure. Actually, most of the relationships which exist between the Design Choices and the Performances Indicators cannot be precisely formalised. In this model, these relations are mainly based on insight of relationships between variables, but not on numerical values of these parameters. Therefore this model aims at providing guidelines to evaluate situations and design opportunities regarding the design of a manufacturing department.

35.2 THEORETICAL BACKGROUND AND AVAILABLE THEORIES

As mentioned previously, we intend to focus on the production process and on the corresponding control structure. For the production process, our approach is based on the production flow analysis developed by J. Burbidge [BUR 79].

For the control of the primary process, we use the approach developed in the GRAI Method [DOU 84], and particularly the decisional aspects in the control structure (see Chapter 18). The horizon and the period of the control decision are taken into account in the model. A part of the illustrated properties comes from the experience of the different partners involved in the ESPRIT basic research project FOF 3143 "Factory Of the Future."

35.3 RELEVANT DC PI RELATIONSHIPS

In this section, we will present the different Design Choices that we use, and the Performance Indicators which are implemented. In describing the Design Choices, we will detail the relations which relate the Design Choices to the Performance Indicators.

Relevant DC PI relationships 429

The Performance indicators are annotated with two initials between brackets (PI). All the other variables which are mentioned are intermediate variables.

In our model, we have considered three Design Choices :

- the type of process,
- the period of decision making,
- the horizon of the decision making.

Six Performance Indicators have been implemented :

- cost of control,
- direct production cost,
- profits,
- customisation level,
- controllability,
- lead-time.

The cost of control represents the costs resulting from the management of the production (control decisions, adjustments, etc.). The direct production cost represents the direct cost of the product achievement, without including the cost of control, the administrative costs. etc.

The customisation level represents the capability and the capacity of the system to provide Customer Driven Products to the market.

The controllability is the capacity of the managers to master, to dominate the production process. This performance indicator is explained in the following section.

35.3.1 Type of process

As mentioned in the first part of this chapter, the main Design Choice we use is the organisation of the primary process which we have called the "type of process". The two extreme situations are "Functional organisation" and "Group Technology organisation". In the following, we are going to describe the consequences of an organisation of the primary process, according to the Group Technology approach. After the production flows analysis, the Group Technology leads to a simplification of the production flows. Such a simplification has at least two main consequences.

First, the Production Leadtime (PI) is shorter. There is less time lost by flowing, and also by waiting and storing. So the Direct Production Costs (PI) are decreased. Less work in process means also that the control is easier. The control decisions are less complex, and this leads to a lower Cost of Control (PI).

From a decisional point of view, GT offers control decentralisation possibilities. The identification of product families allows the identification of several reference processes (groups) which are more or less independent. The coordination of these groups is not too difficult because work orders in these groups are independent. For this reason, the Controllability (PI) of the system is better.

On the other hand, Group Technology leads to less flexibility for providing different kinds of product to the market. If we consider the process as composed of three basic steps like Engineering, Manufacturing and Assembling, Group Technology applied on the two last steps leads to less flexibility for product design. The highest flexibility in that case, is reached if the primary process is completely organised in a functional way. Then any kind of process, within the competence of the company, can be performed by the physical system. The functional organisation of component manufacturing gives a lot of flexibility for the engineering department in designing the product manufacturing process and the product assembling process.

So, Group Technology leads to a standardisation effort, and then, it is to the detriment of the Customer Driven aspects. All these characteristics are illustrated in one Performance Indicator called "customisation level" (PI). The more the organisation is Group Technology oriented (standardisation), the more difficult it is to produce customer driven parts.

The customisation level is increased if the physical system is functionally organised because any kind of product can be manufactured. Regarding the specific Engineering part of the overall process, the question is whether the same principles and the same line of thinking can be followed for Engineering work. A solution for this problem could be to use the human skills required for this knowledge work as the basis for classification. In our model, we consider that an engineering department organised according to function is organised in groups of specialists (experts).
The GT approach corresponds to more polyvalent teams composed of generalists. The first solution (function oriented organisation) leads

to more optimised sub-solutions regarding the specific problem each expert has to solve, but the coordination and the integration of the different sub-solutions is more difficult. So the cost of control (PI) is more important. However, the customisation level (PI) is higher. Thus, we have considered the same kind of relationships for an Engineering department as for a Manufacturing or an Assembling department.

35.3.2 The frequency of the management decisions

The second design choice we have implemented in our model, is the frequency of the control decision. The frequency is determined by the length of time after which the manager adjusts, according to his objectives, the actions to be performed by the process that he controls. For instance, the department manager elaborates a production plan for one week (scheduling). This production plan defines the actions to be taken in order to reach the more global production objectives assigned to this department. As an example, the amount of products to be made per month could be the global objectives.

Every day, he re-considers this action plan according to the results of what has been already done. This is made in order to check that the production results and the new situation (a broken machine for instance) do not threaten this production plan. Thus, the manager is able to take into account the problems that may occur during the execution of the production plan: machine breakdown, quality problems, supplied part missing, etc. Then, it is possible for this manager to adjust this production plan, in order to find a new solution to reach the global objectives that he receives.

In this example the frequency is daily, but obviously, any major problem inside the department may require a real-time reaction.

Through this Design Choice, we aim at illustrating that adjustment frequency has several effects.

The first benefit of a high adjustment frequency (Period = 1/frequency) is an optimisation of the production process. It can be evaluated through the Production lead-time (PI) and the Direct Production Cost (PI) which decrease. Such a situation provides also a better flexibility to the system. It is easier to take into account unforeseen changes. This is interesting especially in Customer Driven production, were uncertainty is predominant. So it increases the customisation level (PI).

However, we also take into account the fact that each adjustment costs money. Each time, the manager has to reconsider his programme on his horizon (this design choice is described in the following section). Regarding the previous design choice, if the control is complex, one adjustment leads to an important effort of information processing. Thus, the Cost of control (PI) is growing.

Furthermore, too much adjustment gives unstability to the controlled system. Then, the production quality decreases so the Direct Production costs (PI), and the Production Lead-time (PI) do also.

35.3.3 The horizon of the management decisions

The horizon of a management decision is the length of time over which the decision extends. This notion of horizon is representative of the scope of the decision which is related to the production cycles of the controlled process.

The horizon of a control decision must be longer than the total cycle of the controlled activities. Otherwise, the manager is "blind" regarding the final steps of the production activities he is programming. For example, if the manager is in charge of a department in which the average lead-time is two weeks, a planning horizon of one week is insufficient. The consequence is that the Controllability (PI) decreases, and the manager tends to be "controlled" by the process.

On the other hand, the Horizon influences the cost of the decision. The longer the Horizon is, the more the decision is complex and the more the amount of information processing becomes important. In consequence, the cost of the decision making is higher, and this increases the Cost of Control (PI). It is also important to consider the reliability of the decision. The more the manager is prospecting in the future, the more the uncertainty grows. Then, there is no more benefit in increasing too much the Horizon. Such a situation is illustrated by the fact that after the threshold where the controllability is maximum, the only effect of an increase in the Horizon is an increase in the Cost of Control (PI).

In our model, these three Design Choices we have just presented are related to each other through their influences on the Performance Indicators we have described. The next part aims at presenting the way the model is implemented.

35.4 DESCRIPTION OF THE IMPLEMENTED MODEL

The model is basically composed of two parts. The main one is the physical system of the department (primary process), and a second block constitutes the control structure.

The model is implemented on STELLA (see Chapter 33). In the following, we describe the way the model is running, and afterwards we will list the main assumptions we have made for this simulation model.

The basis of our model is a simulation of a production process and the control actions on it, at the level of a department. The department receives as an input work orders which have to be performed. These work orders come from the upper level in the hierarchy of the production management system. This level of control is described in the Inter departmental coordination model (Chapter 36).

An amount of capacity is available for the department, with possibilities of internal adjustment (under a certain threshold). The output of the department is the amount of performed work orders. The capacity adjustment mentioned previously is made at each period. The period (Design Choice) is chosen by the model user. At each period, a comparison of the existing situation is made with the "should be" situation. Then, if necessary, a capacity adjustment (under a certain threshold) is made.

However, this real capacity is derived from the theoretical available capacity through a random process. The theoretical available capacity is deduced from a theoretical capacity, because the capacity that the manager expects to use never corresponds to 100% of the theoretical capacity. The random process mentioned above is the illustration of the problems that may occur during the production process (machine breakdown, tools problems,...).

The "should be" situation is represented by the production results which would have been obtained with the theoretical available capacity. In the model, the theoretical situation and the real one are simulated in parallel. This way, it is possible to compare the real situation with the "should be" situation. The theoretical capacity can be adjusted by the upper level at the Aggregate Production Planning level in the Inter-departmental co-ordination model (Chapter 36). This situation is the result of the integration of the Departmental coordination model with the Interdepartmental coordination model.

In order to illustrate the consequences of the main Design Choice, viz. the "type of process", we have also included a specific factor in the capacity adjustment function. This factor aims at changing the amount of capacity adjustment which is possible inside the department.

This factor allows to be taken into account the fact that the ease of changing capacity is related to the type of process. A functional organisation provides more flexibility in terms of capacity utilisation than in a product oriented organisation. Thus, in a functional oriented organisation, it is possible to have more capacity adjustment between activities in the process. Then, it also influences the production lead-time, the cost of control and the production costs.

Our model is implemented for three departments: Engineering department, Manufacturing department and Assembling department. The departments are inter-connected through the Interdepartmental co-ordination model (Chapter 35). The inputs are managed by this last model. The different properties of these departments are interdependent. For instance, the Engineering flexibility is dependent on the value of the Design Choice "type of process" of the two other departments (See remark in 35.3.1).

The costs of production are calculated from the amount of capacity used to perform the work orders and from the time this capacity is used.

The time of the simulation running is around 144 weeks, which represents 3 years. The smallest time unit is the week. This simulation time is independent of the decision horizon. During the simulation, all the uncertainty about the work order arrival, the capacity availability and the capacity adjustment from the upper level (Interdepartmental coordination model), leads to complex situations. The model provides some major indications about the influences of the Design Choices. Such indications are also conditioned by the assumptions we have made in order to model the system. The main assumptions are the following:

- There is no resource conflicts and the capability is not taken into account. The only considered aspect is the capacity.
- The raw materials, the components, the products and the technical information are always available.

These assumptions may be discussed and other Design Choices and Performance Indicators could have been implemented. But then the complexity of the model would have been too high to allow a clear illustration of the described relationships between Design Choices and Performance Indicators.

REFERENCES

1. [BUR 79] BURBIDGE J.L. : *Group Technology in the Engineering Industry.* (Mechanical Engineering Publications Ltd, London, 1979).
2. [DOU 84] DOUMEINGTS G. : *Mèthode GRAI: Mèthode de conception des Systèmes de Production.* (Thése d'etat, Automatique. Université de Bordeaux 1, 1984).

36

Interdepartmental coordination

In Chapter 6 we defined a typology of one-of-a-kind production systems. This typology is based on two questions, namely which activities in the primary process are customer order driven and which investments are customer order independent. For a particular situation, however, it might not always be clear what the main characteristics of the control structure should be. This is due to the fact that a particular company cannot usually be stereotyped in accordance with the typology. Due to the mix of products a company produces, a company can often be characterised as a mixture of two or more types. A machine tool factory, for example, is often characterised as a mixture of a make to order company (MTO for running products) and an engineer to order company (ETO for special products).

In this chapter we will discuss a tool that supports a designer in developing an interdepartmental control structure when moving from an MTO company to an ETO company. The tool focuses on two particular design choices: (1) the level in the control structure at which customer order acceptance (CAO) takes place and (2) the frequency of decision-making at the aggregate production planning (APP) level. Other important design choices that will not be discussed here are the role of the relation between the engineering department on the one hand and manufacturing and assembly on the other.

The discussion of the tool in this chapter will be based on the example of the ETO company described in Chapter 9. The structure of this chapter is as follows. We describe in section 36.1 the old situation of the factory. After a discussion of the design choices and performance indicators in sections 36.2 and 36.3, we will discuss the result of a simulation in section 36.4.

36.1 SITUATION IN THE FACTORY

Traditionally, the company provided a catalogue of more or less standard packaging machines. Parts manufacturing was based on forecasts in this situation and assembly was based on customer orders. Because of product standardisation it was feasible to limit the stock of finished parts. While only assembly was customer order (CO) based, the delivery times were in general relatively short and reliable.

Over the years, however, customers required more specific packaging machines. As a result, parts manufacturing and, more importantly, engineering as well became directly affected by COs. To a certain extent the company was able to produce these customer driven products. Occasionally, however, it caused capacity problems in the engineering department. Moreover, the throughput times of parts manufacturing became increasingly unstable and overall lead times increased. More specifically, there were some problems with respect to customer order acceptance. Customer orders with short delivery times were accepted, while the work in progress was high. It was obvious that promised due dates could not be met. Moreover, large differences in customer order size led to instability concerning capacity requirement and work load, which were based on forecasts. Engineering and manufacturing especially were affected. These effects worsened over time. Furthermore, the Sales Division was trying to find a more competitive advantage for the company. They found that CO driven production of customer driven products would be the solution. In addition, a reduction and stabilisation of delivery times would be necessary, while extra costs should be limited.

For this reason, the company felt the need to adapt its control structure. The old control structure is schematised in figure 36.1. Interesting points are: (1) the engineering department is controlled by the aggregate production plan (APP), (2) APP has a low frequency, (3) customer order acceptance takes place at the factory coordination (FC) level.

36.2 DESIGN CHOICES

A new control structure should avoid the problems mentioned above, and should at the same time stabilise customer order lead times and minimise costs. Finally, the rate of rejected customer orders due to lack of available capacity should decrease considerably.

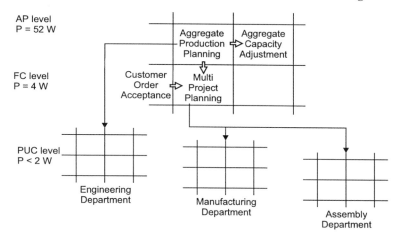

Fig. 36.1 Control architecture (grid)

As stated above, the computerised model focuses on two design choices concerning the control structure, namely (1) the level of customer order acceptance, and (2) the frequency of APP decision-making. Both design choices are discussed in the following sub-sections.

36.2.1 Level at which Customer Order Acceptance takes place

Customer orders can be accepted at the aggregate planning (AP) level or at the factory coordination (FC) level. We assume that substantial capacity adjustment is only possible at AP level. At the FC level, COs can be accepted on a relatively short term, and due dates can be based on the actual state of the production system. In the case of high work loads, however, COs will not be accepted, or only with long due dates. This could lead to high rejection rates. Because of the limited possibilities of capacity adjustment at FC level, this could at the same time result in an unstable system.

On the other hand, capacity adjustment is also possible at the AP level. Customer order acceptance at the AP level will result in low rejection rates, especially if it is combined with capacity adjustment. However, the frequency of customer order acceptance will decrease, which has negative effects. This dilemma can be studied by the

440   *Interdepartmental coordination*

computerised model to be discussed here, the interdepartmental coordination model. The inputs for the simulation are:

- the level of COA
- the frequency of AP decision-making, and
- the size, variance and uncertainty in COs.

The outputs of the simulation are:

- the rejection rate of orders
- the resulting lead times and
- the costs of production and decision making.

Four situations might occur, two of which are interesting in the specific case:

1. When there are only a small number of large COs which are all about equal in size (low uncertainty and variance), COA at FC will result in a relatively stable system with frequent capacity adjustments at the AP level. Sometimes COs will be rejected due to capacity limits. COA at AP level will result in the acceptance of all COs. The system will also be more stable because capacity adjustment is better in tune with the COA. The decision-making costs will increase, however, due to larger overheads.
2. When there are a small number of large COs and there is a large variance and uncertainty in size, COA at FC will result in an unstable system. There will often be capacity adjustments, both at the AP and the FC level, and there will be a relatively large number of CO rejections due to capacity limits. COA at AP level will in this case result in a more stable system because of the better tuning of capacity adjustments, relatively large in size and number, at APP level. Problems may occur, however, due to a relatively low decision-making frequency at AP level.

The other two situations are:

3. there are a large number of small COs which are all about equal in size (low uncertainty and variance)
4. there are a large number of COs and there is a large variance and uncertainty in size. In the company used as an example, however, we are most interested in the first two cases.

36.2.2 Frequency of APP decision making

The frequency of decision making at the aggregate planning level is an important factor for both customer order acceptance (if it takes place at this level) and long term capacity adjustment. On the one hand, high frequency of decision making results in flexibility vis-à-vis the market. On the other hand, the costs of this high frequency of decision making increase very rapidly. Therefore, an optimum frequency needs to be found. Using the tool, different frequencies can be simulated.

36.2.3 Performance Indicators

The performance indicators (PIs) of this model are:

for each department:

- work in progress (WIP)
- backlog
- available capacity
- production costs
- decision-making costs
- average lead time
- number of work orders per week.

for the whole factory:

- total costs
- average of the total lead time
- COA rejection rate
- total production volume
- average costs per product.

The following design choice options (alternatives) will be evaluated:

level \ period	8	12	16	52
COA at APP				
COA at FC				

A number of performance indicators (PIs) can be evaluated for each cell in this matrix.

36.3 EVALUATION

Based on the simulations, two suggestions for improvement can be made. Firstly, the question of whether to accept customer orders at the aggregate level or at the FC level needs to be discussed. Based on the simulation results, it is recommended that a COA procedure be developed in which large orders are accepted at the aggregate level (large in terms of capacity requirement or OKP-ness) and small orders are accepted at the FC level. Secondly, there is a need for short term feedback on the availability of capacity. Therefore, the frequency of aggregate decision-making needs to increase.

Hence, the tool indicated possible changes in the control architecture that are required to adapt to the market requirement of producing more customer defined products. The model did not, however, simulate the organisation consequences of these changes in terms of human capacity, for example. A new procedure in customer order acceptance, for example, means that other people are involved, meetings have to take place at different frequencies, other authorities and responsibilities are required, etc. A change in APP decision-making frequency would also affect the agenda of this meeting. Other short term issues would be discussed. A more detailed analysis of the situation needs to be made to discover these effects. This can be accomplished by using simulation models that describe these decision-making functions in more detail. For example, some of the consequences for departmental control can be simulated by the Departmental Coordination model.

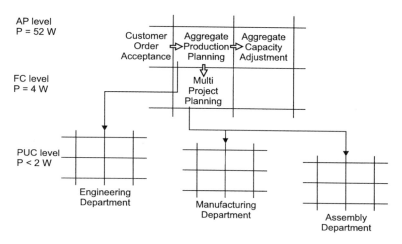

Fig. 36.2 New control architecture

37

XBE in Design of Customer Driven Manufacturing Systems

37.1 INTRODUCTION

The design of production and production management systems is an area in which the computer support available has differed in different kinds of simulation systems. Some tools that can be regarded as decision support tools do exist but, typically, the level of aggregation is rather low, as it is with shop floor simulation systems.

Particularly at the strategic level, where general managerial principles are chosen, factory re-design is an area with a high level of abstraction and complexity and a need for creativity. An example of such a planning task is the design of a new production management concept (Riis 1990a,b). To be successful, a multitude of different aspects (layout, machine types, workforce remuneration) and a multitude of different processes (engineering design, parts production, assembly, purchasing) need to be taken into account. This chapter describes the rationale of such design activities and how it led to the idea of an expert support system. For the application area, a hypermedia based toolbox was built, on top of which an application is currently running as a demonstration prototype. The central building blocks of the system are described, as is the use of the demonstration application.

37.2 PROBLEM AREA

The need for continuous improvement forces manufacturing systems to make re-design efforts continuously. The reason for this can usually be classified as a problem or a potential for improvement. Typically, three steps can be distinguished in this kind of engineering process (Eloranta 1991):

1. Evaluate the current situation
2. Specify goals for change
3. Specify a new design to meet the goal

A typical feature of the design problems we are addressing in this chapter is a high level of aggregation. To give an example, for the rough design of a production management system the third step can be described as choosing the operating principles under the existing structural design and changing the structure and organisation of the target system to be able to apply new operating principles. The goals of these change actions vary, but nowadays reducing lead times, inventory levels and improving delivery performance and customer service are most often important.

Analysing a situation and specifying new goals is mostly an analytical task in which methodological means can be applied in a standard form. Making the right design choices (and not making the wrong ones) is a difficult task, because there are no two factories that are exactly alike. To avoid re-inventing the wheel at great expense companies often rely on consultants, who have both experience in and references from re-design efforts.

37.3 THEORETICAL BACKGROUND

The toolbox we describe here is called XBE (eXpert system By Examples) and the application we have built with the toolbox is called XBE-BET (BET for Business Engineering Tool). The XBE is based on the situation-ends-means reasoning in design problems. The picture below illustrates the idea.

The central idea is to mimic and support the actions of a consultant in the process of analysing, remembering former cases and combining different pieces to reach a new solution. Referring to the picture below, the inferencing process is as follows. Production Management System Consultant XXX visits a factory in trouble. New machinery has been installed with a completely new layout and group technology organisation. However, there have been no positive results from these change actions. After interviewing management and some shop floor employees consultant XXX comes to the conclusion that the line of action taken was right. Nevertheless, something is missing. Luckily, consultant XXX remembers clearly the case of factory YZ, where in a similar situation one more change was added: a new wage system with incentives. This incentive system did not actually create any additional expense, so perhaps something similar would work. The incentive system was implemented with the necessary changes correlating to the type of work in the problem situation. It seems to have done the trick: within two months half of the productivity expectations have already been achieved.

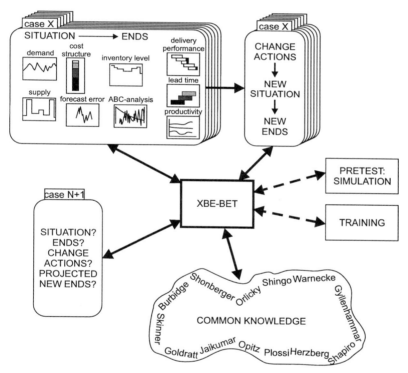

Fig. 37.1 Basic idea of XBE

The more complicated the situation, the harder it gets to invent all the possible or necessary change actions needed in a factory to keep it competitive. This problem area has been addressed using various techniques. Mathematical models have failed, perhaps because the complexity of the problem would lead to an enormous model if described in detail. The mainstreams of AI research, rule based and model based approaches, were tested as well and were found very difficult to apply (Eloranta et al., 1990). In XBE a fairly new approach, case based reasoning, is applied. This approach has the same theoretical basis as machine learning (see e.g. Michalski et al., 1991), but it is currently more applicable to complex problems. A more detailed description of the modelling approach in XBE can be found in (Pankakoski et al., 1991).

37.4 THE TOOLBOX

At the start of the research work there was only a vague idea of the nature of the tool that was needed. As the literature on case based research focuses only on less complex problem domains than manufacturing systems design, we decided to concentrate on prototyping with HyperCard. It soon became clear that we really needed a toolbox before we would be able to give a full-scale demonstration of the central ideas. To ensure that the abundance of information on a manufacturing system can be described in the tool, it is necessary to have hypermedia capabilities to enrich numbers with pictures, free-form text and various hypermedia links. The requirements for the system are the following:

- A dialogue design tool to support numbers, text and pictures
- A database for numbers, text and pictures
- Multiwindowing capability to support the idea of considering a design target from multiple aspects
- Hypermedia links to record associations between objects
- Graphical toolbox to present numerical information graphically

Inferencing mechanisms to compare cases by a CBR-approach (Pankakoski et al., 1991):

- trend (of a time series)
- correlation (of a time series)
- Euclidian distance (of a weighted vector)
- keywords.

A real thrust to the development of the prototype was given when a fully HyperCard compatible database, HyperHit_, was found. Hyper-

Card allows the implementation of databases, but the possibilities are limited. HyperHit_, on the other hand, allows the database structure to be changed on the run, which is a central capability for the toolbox. This capability allowed us to fully automate the creation of a database structure with the BOA and semi-automate the design of dialogue screens.

As a toolbox, XBE has already reached the level where the gains from further automation of the application development process can be considered too expensive. Customisation of the available procedures is an easy task to accomplish for anyone with a reasonable knowledge of HyperTalk, the English-like programming language of the HyperCard system.

In the FOF project (Hamacher and Hirsch, 1991), in which the results described in this chapter were mainly achieved, two 'models' were implemented using XBE: the Economic model (XBE-BET) and the Simultaneous Engineering model. The Economic model focuses on choosing the general operating principles for running a customer driven manufacturing (CDM) system. The Simultaneous Engineering model focuses on the development of the engineering process, especially using the means generally known as simultaneous or concurrent engineering.

37.5 DESCRIPTION OF THE MODEL

This section describes the structure and the use of XBE-BET. In the terminology of CIM-OSA (AMICE 89), XBE-BET is a system for describing and storing particular models of manufacturing companies. However, XBE-BET is not a simulation model. Inside XBE-BET, the user cannot test the effects of changes, e.g. a new layout. If simulation is needed, the case information could be extracted from XBE-BET and fed into a simulation tool.

A case description contains information on the case situation, the ends (goals of the development project) and the means chosen to achieve the ends. The corresponding theory used in XBE-BET is discussed in Chapter 2. In XBE-BET, the business process structure of CIM-OSA (ESPRIT-AMICE) has been adopted to avoid the need to describe the whole system all the time and to be able to describe different companies in a uniform structure. A special effort has been directed towards the product engineering process, which is the heart of customer driven manufacturing. Business processes that have been described are the following:

- Product Engineering
- Production Management
- Parts Production
- Assembly
- Purchasing

Activity: [A1] Analyse Performance Data
With XBE, a re-design project usually starts with a set of quantitative analyses that start from a very rough level of information, e.g. turnover, net profit, total throughput time, etc. The purpose of these analyses is to help the user to select the most important business processes, i.e. those that are most relevant for the success of the case company and have potential for improvement. These business processes will then be further analysed and described to find feasible design choices.

Activity: [A2] Describe Target Area
A set of presentation tools has been built and used in XBE-BET. With the aid of these tools the user can easily build new screens for the application. Types of information that can be presented with these screens include:

- time series (yearly costs, monthly sales,...); numerical and graphical presentation
- quantitative values
- quantitative scales (average number of level in BOM)
- qualitative values
- qualitative scales
- sets of values (i.e. 'choose one of these')
- plain text
- pictures
- links (e.g. hypermedia-like).

The user will be able to describe a manufacturing system with a hierarchical structure of business processes. For each business process, variables describing at least the current values for frame conditions, design choices and performance indicators are used to describe the "as is" situation. Additional descriptive information can be added to enable the user to describe situations in greater detail.

The user can freely choose the level of abstraction in analysing a case factory. Sometimes just the rough level performance information and a general company/factory level description of the company is enough to use XBE-BET. On the other hand, detailed descriptions of business processes are supported as well.

Activity: [A3] Search for similar cases
Indexing and the retrieval of existing cases from the case bank is the 'inferencing' XBE provides for the user. This procedure is called matching. From a theoretical point of view, matching should be based on variables describing the frame conditions of cases and business processes. On the other hand, like human memory, cases (i.e. experiences) should be linked to each other in various ways. Only 'keywords' can be used without a case. All other indexes are relative, i.e. the current case is used as a basis for comparisons.

The user can choose the level and type of information s/he uses for searching. Someone might be interested in searching cases with a similar kind of inventory structure, or the user could ask very plainly: 'In which cases has throughput time in assembly been improved?'

The user can use several types of search on a set of cases to reduce the number of cases to be browsed through. For example, a first criterion for the user could be 'companies doing pure CDM business'. A second criterion would be 'companies of our size' and the third 'companies with similar assembly function/layout/organisation'.

Activity: [A4] Browse through cases
After obtaining a selection of companies that could potentially benefit the user, it is time to check the validity of the cases. The act of browsing needs to answer two questions:

1. What change actions have been taken?
2. What kinds of effect have they had?

In XBE-BET, cases are described in 'snapshots', i.e. 'as is' descriptions in succinct moments of time. These intervals are called phases in XBE-BET, to recall the development cycle of factories. A sensible timescale for these intervals is at least one year, as the shortest time to implement any real changes in a factory seems to be at least six months.

XBE-BET includes two possibilities for describing design choices (question 1). The first one is very close to the strict theoretical approach. Values of design choices at the moment t vs values at the moment t+1 is the basic approach. In XBE-BET, there is a slightly vaguer (and more dynamic) possibility for describing design choices: the user can describe change actions, i.e. what will be changed, why (global and local performance indicators), where (business process) and how. The format is fairly free, but the answers to the questions mentioned must be given.

In a real case, there is only one possibility for finding out the effects (question 2): to compare the performance of the case at two different times. The first moment in time would be, for example, the initial first-time analysis and the second moment a performance analysis made a year or two later.

Activity: [A5] Choose suitable design choices

What combination of changes from those cases the user considers to be relevant for the current case is feasible? This difficult task has been left to the user. Adapting/modifying the solutions found from the relevant cases for the current case is not a trivial task. A much deeper understanding of what factors can be changed and how they can be changed is required before this step (to create a development plan) of the case based reasoning process in XBE-BET can be automated, even partially.

After choosing suitable design choices, the user draws up a development plan (or several). This can be done in XBE-BET. Usually, the user draws up several development plans. The procedure could be as follows. The user has found a suitable set of design choices to develop the factory as a whole, but these design choices do not give enough boost for his/her engineering process, which s/he considers crucial. Therefore, s/he makes a new loop with XBE-BET: describe the engineering process, search for solutions, etc. The new loop finds enough relevant cases in which the engineering process has been successfully developed, so s/he can draw up a special plan for the engineering process.

Activity: [A6] Update case bank

XBE-BET allows the use of multiple databases, so there could be one database for CDM production, one for mass production, etc. The hierarchical structure mentioned in [A1] is specific for every database. So the logic of the system can at least be retrieved in part from the database. As XBE is meant to be used by several people simultaneously, a maintenance tool has been built. With this tool the user can move whole cases or just partial information from one database to another.

Information we think is common to all the databases is stored in a central database. This database contains mainly structural data, e.g. currencies.

As the user fills in data for his/her case, it is actually handled as any other case in the databank. The data is updated when the user exits/changes different screens.

CONCLUSION

Manufacturing systems design is a complex task and the long term decisions/plans in particular involve a lot of intuition together with qualitative and quantitative information. To support this kind of decision-making with computers, it is suggested that case based reasoning is a suitable methodology. A hypermedia based toolbox is presented, as well as the use of an application (XBE-BET) built on top of the toolbox.

At present, the prototype system is well capable of presenting the functionality of the tool and the central ideas of the approach. The case bank in the prototype currently contains about a dozen cases. The prototype is already being tested in small consultancy projects being executed by the research group. In the future, we think case based reasoning will find a lot of similar, successful application areas and happy users.

REFERENCES

1. Eloranta, E.: XBE-BET, *A Drawing Model in the FOF Demonstrator*, Internal Working Paper, ESPRIT II Basic Research Action No. 3143, February 1991.
2. Eloranta, E., M. Syrjänen, S. Törmä: "A Knowledge-based Tool for Manufacturing Systems Design", *Computer Integrated Manufacturing Systems*, Vol. 3, No. 3, 1990.
3. ESPRIT Consortium AMICE (Eds): *Open System Architecture for CIM*, Springer-Verlag 1989.
4. Hamacher, B., B. Hirsch: Workpackage Report 6 Report, Cyclic Rapid Prototyping, *ESPRIT II Basic Research Action* No. 3143, FOF, October 1991.
5. Michalski, R.S., J.G. Carbonell, T.M. Mitchell (Eds.): *Machine Learning: An artificial Intelligence Approach*, Tioga Press, California, 1991.

6. Pankakoski, J., E. Eloranta, M. Luhtala, J. Nikkola: "Applying Case-Based Reasoning to Manufacturing Systems Design", *Computer Integrated Manufacturing Systems Journal*, November 1991.
7. Riis, J. O.: "The use of the production management concept in the design of production management systems", *Production Planning & Control*, Vol. 1, No. 1, 1990a.
8. Riis, J. O., H. Mikkelsen: "Planning and managing production management improvement projects", *Production Planning & Control*, Vol. 1, No. 2, 1990b.

Index

Page numbers appearing in **bold** refer to figures and page numbers appearing in *italic* refer to tables.

Age distribution 408–10
Aggregate Production Planning (APP) 244–5, 251, 252, 266, 268–9, 271–2, 289, 433, 437–42
Algorithms 298
Analytic models 15
Anonymous products 69–70
Apple Macintosh platform 402, 413
Architecture 327–8, 386
Area Coordinators (ACs) 91
Assemble-to-Order 62, 64, 69, 105–18, 265,
   see also Medicom
Assembly 64, 206–7, 382
Automatically Guided Vehicle system (AGV) 393
Available-to-Promise (ATP) 261, 264

Batches 35, 336
Bill of Material (BOM) 49–50, 70, 80, 100, 144, 194, 197, 260, 261, 263–4, 282, 321, 327, 338
   variant 321, 322
BIRK-92 309, 311
Boolean operators 328
Bottling Technology 91
Bout-93 310
Business
   evaluation 286–7
   information systems 333–56, 335
   Process (Re)Engineering 189

Capabilities/capacities 418, **419**–20
Capability, nature of the required 169–70

Capacity
   analysis 388
   planning 65, **293**, 294–6
   selling 65
   structure 250–1
Capital goods 38, 305
Cases, introduction to 85–7
Change Management 361–2, 366–8
Changeover schedules 246–7, 251
Communicating order information 206
Component manufacturing 66, 382
Computer
   Integrated Manufacturing (CIM) 194, 388, 449
   Supported Cooperative Work (CSCW) 307
Computer-Aided
   design (CAD) 143, 151, 194, 196, 308, 319, 359, 373
   engineering (CAE) 151
   production management packages 224–5, 298
Concurrent
   engineering 219, 306, 313
   mental modelling 218, **219**
Control responsibilities **76**–7, 170, 282
COSIMA project 386
Cost of control 163, 430, 432
CRP program 350
Customer
   after-sales service **304**, 308
   orientation 22–3
   supplier relations 79, 303–14

456 Index

Customer *contd*
  views **199–200**, 199–200, 202–4, 320
Customer driven
  businesses 62, 63, 346
  engineering 78, 159–75, 281–8
    Information Technology (IT) 375–84
  manufacturing (CDM) 23, 33–4, **39–40**, **42**, 78–9, 223–4, 233–7, 301
    customer relations 306
    document management 357–72, **360**, 372
    IT 317–18, 333–56, 449, 451–2
    production 177–92
    typology 59–73, 231, 241
    value–adding customer 23, 27–30
  production systems 210–1
  products 429
  work-flow-oriented manufacturing 67
Customer Order Decoupling Point (CODP) 59–60, 69, 151–2, 249–50, 254, 321–2, **323**
Customer-integrated, globally distributed order processing **307**
Customer-order-driven systems 219–20
Customer-order-independent information 72
Customer–orders (COs) 69, 127, **242**, 275, 283, 352, 376, 385, 438, 440
  acceptance (COA) 200, 206, 261, 275, 437, 439–42
  assembly 64
Customer-orders-specific information 72
Customer-specific products 309

Daimler Benz 37
Data 390
  analysis **391**
  base management systems (DBMS) **372**, 373
  format Odette 311

  gradually emerging **349**, 352–3
  management 317, 357–8
  models 376–84, **377–9**, 383–**4**
    generic bill-of-material 319–22, 320, 328
  structure diagram **340**, 348–51, **349**
Decisions 55–7, **56**, 223–37, **278**, 279
  management 431–2
  structure 274–9, **278**, 283–5, **284**, 290, 291
  support 70, 353–5, **355**, 393
Decomposition criteria 227–9, **227**
Delivery dates 206, 275, 290, 296
DELTA 423
Departmental coordination model 398, 427–35, **427**, 433, 442
Design 159
  consideration, manufacturing technology 186–8
  principles, transformation process 165
  problems **445–6**
  requirements, cost control 163
  rules 326–31, **326–8**
Design Choice – Performance Indicator (DC–PI) 9, 15, 19, 428
  modelling 10–12, 15, 17, 19, **19**
  network 12–13, 14, 405
Design Choices (DCs) 11, 18, 406, 421–3, **424**, 427, 429, 432–5, 438, **439**, 446
  intelligent storage and retrieval 401–6
Design-oriented approach 9–19, 76–8
Detail Design 174, 271, 276–7, 284, 288, 375, 381–2
  orders **355**, 376
Detailed drawings 10
Development procedures 29
DIMUN (Distributed International Manufacturing Using Existing and Developing Public Networks) 306–8, 307
Dispatching 296–300, **300**, 385–6, 388–90

intelligent storage and retrieval 401–6
see also Design Choice – Performance Indicator (DC–IP)
Physical
  flows 224
  system 225
PIMS data base 23, 25–6
Planning 68, 243, 387
PMC strategy 286
Primary Process 78–9
Process Planning 92, 94, 174, 271, 282
Producer-supplier cooperation 309, 310
Product-oriented
  companies 60–2, 72, 86, 105–18
  data structures 380
  make-to-stock production 336–43
Production **151**
  activity control 289–301, **289**, **291**, 293–4, 299–300, 386
  control 95, 241–2, 248–9, 271–4, 292, 300, 373, 427
    design principles 267–9
    engineer-to-order firms 267–79
    engineering process **162**, 171–2
    four key decisions **278**, 279
    system 79, 165, 241–56, **242**, 244, 252–3, 255
  costs, direct 430–1
  customer driven manufacturing 177–92
  documentation 282
  equipment designers 11
  Flow Analysis (PFA) 177–8, 184–6, 189
  leadtime 430–2
  management 68, 237
    systems (PMS) 226, **227**, **229**, 233–4, 386, 446, **447**
    typology 63–8, **65**, 151, 152, 437
  methods, environmental protective 35
  monitor 391–2, *391*
  phases 165–8, 174
  Planning and Control (PPC) 148–50, 298
    process structure 249–51
    and sales **242**, 245–7, 269, 273–5
    scheduling 387
    Schw-92 309
    systems 33–4, 33–5
      designers 10–11
      economic and technical challenges 35–7, **36**
      prototyping capabilities **39**
  Unit Control (PUC) 126, 166, 243, 247–8, 268, 270, 274, 277–8, **289**, 296, 337, 346, 375
    architecture 390
    decision structure 290, **291**
    information system 385–94
  Units (PU) 165–6, 169–70, 173–4, 178–80, 183–4, 250–1, 255, 274, 276, 278, 282, 297, 347
    control architecture 290, **443**
    establishing 270–1
    self-containedness 249
    work loads 291–2
  volume 67
Production management 65–8
Production-oriented view 51
Products
  data 195, **384**
    management 373
  description 159
  development 309–13, 310, 312, 326–7
  documentation 317
  engineering 67, 159, 174, 271, 276, 285, 287, 375, 377
  orders 355, 376, 381
  families 324–5, **324**
    documentation 358–9
    point of view 321
    structure **201**
  flow lines 224
  groups 206
  modelling 193–208
  models 195–7
  strategy 67

Products *cont*
  structure 167–9
  variants 64, **196**, 197–205, 321–2, 327–30
Project management 68
Prototypes 37

Quality 24–5, 188
  assurance procedures 359
  circles 29
  control, Medicom 262
  Function Deployment (QFD) 305
  monitor 392–3
  problems 67
Quotations
  due dates **242**, 245–7
  evaluating and selecting requests 285–7
  issuing 287–8, 375, **379**
  phases 93–4, **162**, 163–4, 375, 380
  team 287

R-graph 53–**54**, 56
Rapid Prototyping Technologies (RPTs) 38
Raw materials monitor 392, *392*
Reference
  engineering review 378
  product modules 377
  sub functions 377, 384
Relationship models **14**
REMBRANDT (REference Model BRowser ANd Design Tool) 397–8, 402–6, 425
Research and Development (R&D) 218
  cost curves **36**, 36–37
Resource
  capacity bottlenecks 167, 250, 252
  management
    basic requirements 211–14
    challenges 216–20
  units, complex 216–17
Resource-oriented
  companies 60–3
  manufacturing 137–52

Resources
  durable and non–durable resources 57, 211–12
  human 57, 90, **91**, 350–1, 417
    and non-human resources 57, 212, 213
  properties 211–14
  structure 180–4
  tangible and non-tangible 57, 213
  visible and invisible 213–14
  and work-flow 52, 350, 373
ROI 25–6
Routings 339, 344

Sales 90, 93, 121, 123, 205–6, 438
  after-sales service **304**, 308
  engineering
    and materials management 288
    and production 90, 173–4
  pre-sales 304–5
  and system operations 205–6
Scheduling 385–8
  capacity allocation 294
Selective Laser Sintering (SLS) 38
Service organisations 63
Shipbuilding 87, 238–9
Ships and maritime products
  company 137–52
  administration 139
  CAD-system 143, 151
  design **141**, 143–4, 149–50
  dock assembly 146–7
  engineering phase 140, 142–4
  inter-organisational production **147**, 148
  internal organization 139–40
  manufacturing 139, **140**, 144–7, **145**, 150
  operations planning **144**
  pipe manufacturing **145**, 146
  problem areas 151–2
  Production Planning and Control (PPC) 148–50
  products **137–8**
  project definition **140**
  quotation planning 141–2

Ships and maritime products
  sales 139
  steel sections 142–3, **145**, 146
  transformation process (work-flow view) 140–8
  WOST concept: Workshop-Oriented Ship Production Technology 139, 147
Shop floor
  lay-out 190
  management 80, 177, 279, 385, 389
Simulation model, technical structure 413–16, **414–15**
Simultaneous Engineering (SE) 305, 449
Software
  and hardware platform 402–3
  packages for business information systems 333–56
Speed 26
STandard for the Exchange of Product model data (STEP) 195
Standard Lead Time (SLT) 337, 339
State-dependent
  data model **379**, 380–4, **383–4**
  data structure diagram for MRP 342–3, **342**, **354**, 380
  transaction processing systems 335–6, 341–2, 351
State-independent
  data model 376–80, **377–8**, 382–4, **383–4**
  data structure diagram for MRP 339–41, **340**
  PUC–workflow data 378–80, **378**
  transaction processing systems 337, 345–8, *347–8*
STELLA, Systems Dynamics model 413, 416, 433
Stereolithography (SLA) 38
Sub-order assignment and outsourcing 275–7, 284, 382
Subcontractors 309
Systematic Handling Analysis (SHA) 190
Systematic Layout Planning (SLP) 190

Systems 200, 413
  parameter 201, *203*

Tayloristic production systems 407
Teamwork 29, 40, 286–7
Technical
  Documentation & Administration 92
  engineering 100
  evaluation 286, 380–1
  Information Management Systems (TIMS) 311
  sources 122
Tendering and engineering 281–8
Third World countries 34
Time-to-market 37
Transaction processings 69–70
Transformation processes 89, 93–7, 111–13, 122, 140–8, 165, 269–71

UN, industrial statistics 26
Uncertainty, reducing 167
Uniplex multiview unit 307–8

Value-adding customer driven manufacturing 23, 27–30
VDA-IS interface 311
Verein Deutscher Ingenieure (VDI 2221) 161, **162**
Vertical decomposition 228–9

Walras
  decision description 55–7, **56**
  production model 47–57, **48**, 50, 53, **57**
  resource description 53–5, **54**
  work-flow description 49–53, **53**
Walrasian model 227, 248
Work
  loads 290–4, **293–4**
  orders 294–6, 353
  release 277–8, 290, 292, 296–7
  sequencing 278–9
  systems 214, 215

Work-flow
  management 80, 364–5, 370–1
  and resources 52, 350, 373
Work-flow-oriented
  companies 61–3
  data structures 380
  make-to-order firms 241–56, **242**, **252–3**, **255**

production management 68
  view 51
Work-In-Next-Queue (WINQ) 299
Working groups 418–20, **419–20**

XBE-BET (Expert System by Examples) 397, 399, 446–54, **447**

LANCHESTER LIBRARY

3 8001 00201 8541

/59.00

LANCHESTER LIBRARY, Coventry University
Gosford Street, Coventry CV1 5DD Telephone 024 7688 7555

12 FEB 2009

02 MAR 2010

This book is due to be returned not later than the date and time stamped above. Fines are charged on overdue books